本 书 编 委 会 \ 编

中国林业出版社

域 上

中 国 室 内 设 计 年 鉴

INTERIOR DESIGN YEARBOOK OF CHINA

中国林业出版社
China Forestry Publishing House

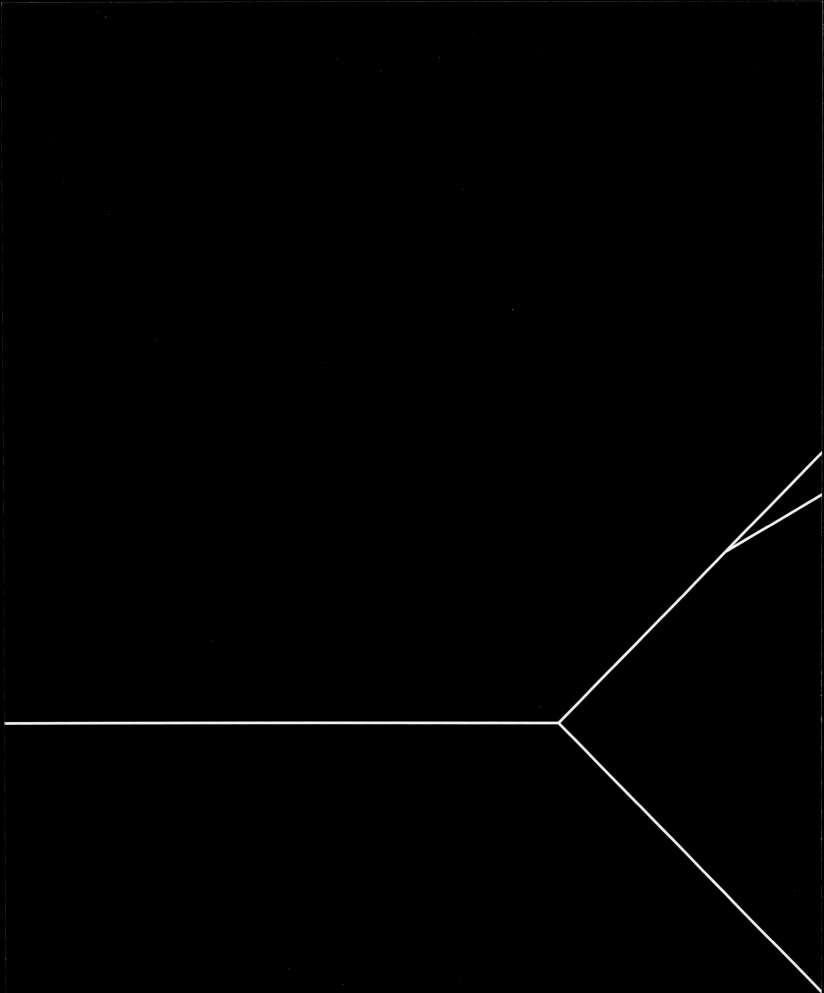

设计版图！

近年来，中国的室内设计取得了长足的进步，本土的设计力量逐渐浮出水面，这让我们看到了中国设计崛起的可能性，但某种程度上，境外的设计师力量仍然占有某些优势。中国是全世界范围内设计最为蓬勃发展的设计热土，每年这里都会涌现出无数的作品，这催生了设计出版行业的繁荣，这些图书及杂志为我们记录了当代中国设计的发展动态，是我们了解当下中国设计的重要路径。坊间每年都会出版几本中国室内的年鉴类书籍，他们是我们了解本年度室内设计发展的重要窗口，但是这些室内年鉴在编选的时候，侧重于本土设计力量，而且是按照项目类型来编辑的。因此，我们尝试以一种新的方式来解读当下中国设计的新发展。我们没有按照项目的类型来编辑设计作品，而是按照行政区域来划分，而且我们的视野不仅停留在本土的设计力量，而是关注一切在中国的设计力量。

中国幅员辽阔，地区间社会、文化及经济发展存在非常大的差异，相应地，设计的发展状态也不太一样，尝试去勾勒一个大致的中国当下设计版图也许有助于我们更加理性认识当下中国设计的发展及状态。大致来说，目前大陆地区，设计最为活跃、最为繁荣的区域是华南地区、华东地区、华北地区，华南地区以深圳、广州等为代表，华东地区以上海、苏州、杭州、厦门、福州等地为代表，而华北地区则以北京、天津为代表。当然，勾勒中国的设计版图不可能少了香港、台湾。这些区域涌现出的设计作品构成了我们本书的主体。这些区域之外，华中地区、东北地区也有一些优秀的设计作品呈现，但是西北地区及西南地区的设计作品较少进入我们的视野。一方面，这也许受我们自身视野所限，但另一方面也显示了目前西南及西北区域的设计力量仍然比较薄弱。

虽然我们尽力勾勒当下中国清晰的设计版图，但这异常困难。当下中国每年涌现的优秀作品实在不胜其数，由于能力的限制，我们在编选的时候，难免会遗落大量的杰作。但是我们希望读者可以从我们选择的作品中，看到本年度中国当代室内设计的发展动态，窥一斑而见全豹。上海、北京、香港、台湾、深圳等地是传统的设计活跃城市，在这些地方不仅仅本土设计力量强劲，而且国际性的设计力量也非常强大。在华东区域，厦门、福州是非常值得一提的城市，这两个城市的设计力量近年来非常活跃，在国际国内都获得了非常多的奖项，他们浮出水面是中国设计取得长足进步的结果。华北区域的天津虽然也非常值得关注，近年来也涌现了非常多的优秀作品，但这些作品的设计师不是本土的设计力量，而是来自境外或者上海、北京、深圳等地设计力量。这显示当地设计力量的发展得益于政策的推动，但仅仅只有政策的推动并不足以让当地的设计力量成长，还需要当地设计力量自我学习及反思。西南区域的重庆，近来年也成为了各种设计力量角逐的战场，据说当地每年要兴建十几个五星级酒店，由此可见当地设计市场的火爆，但是如此火爆的设计市场并没有孕育出对称的设计作品。这其中的缘由值得我们反思。香港设计对内地特别是深圳设计的影响早已成为共识，但是一直以来，我们虽然也非常关注台湾的设计力量及其对大陆设计的影响，但我认为，我们的关注仍然不足。台湾地区的设计既有中国传统文化的影响但又非常国际化，在我看来，这是中国当代设计自成一格的不二法门。因此，在本书中，我们特别将台湾地域辟为一章，以彰显其重要性，也希望引起业内对台湾设计的进一步认识。

In recent years, China whose local design strength has been revealed has made considerable progress in interior design, which makes us see the possibility of rise of Chinese design, but to some extent, oversea designers force still occupy some advantages. China is the most booming design hot spot worldwide and numerous works emerge every year here, which has given birth to prosperity of design and publishing industry recording development trend of contemporary Chinese design by books and magazines for us as the important path that we learn about contemporary Chinese design. Several indoor Yearbooks newly published every year are known as an important window to understand interior design development of the current year, but these indoor Yearbooks when compiled have been focused on local design strength and edited in accordance with the project type. Therefore, we try a new way to interpret the new development of contemporary Chinese design. Instead of project type, we have used administrative region division to edit design works, and in addition, we focus our vision not only merely on the local design force but also on all design strengths in China.

Many social, cultural and economic development differences between regions of China as a large country, accordingly result in various development design states, so that how to outline a general contemporary design layout in China may help us to more rationally understand development and state of contemporary Chinese design. Generally speaking, the most active and most prosperous design regions in the mainland are South China represented by Shenzhen, Guangzhou, etc, East China represented by Shanghai, Suzhou, Hangzhou, Xiamen, Fuzhou, etc, North China represented by Beijing, Tianjin, etc. Of course, it is impossible to outline China's design layout without Hong Kong and Taiwan whose design works emerged to constitute the main body of our book. Besides the above mentioned areas, Central China and Northeast also present some excellent design work, but the design work from Northwest and Southwest areas is beyond our field of vision. On the one hand, our own vision may be limited, but on the other hand, also it shows that design strength of Southwest and Northwest regions remains relatively weak.

Although we make every effort to outline Chinese current clear design layout, it is extremely difficult. At the moment, China's outstanding works annually emerge is really numerous, due to capacity constraints, during compilation, we could inevitably neglected a lot of masterpieces. But we wish our readers to see the development state of Chinese contemporary interior design of the current year from 100 pieces of works that we have chosen, jumping into conclusion through fragment observation. Traditionally, design activities are vibrant in such cities as Shanghai, Beijing, Hong Kong, Taiwan, Shenzhen, etc, whose not only local but also international design strength is very powerful. In the region of East China, Xiamen, Fuzhou is very worth mentioning, whose design strength are very active in recent years reputed by lot of international and domestic awards, resulting from great progress in Chinese design. Tianjin of North China is also very worthy of attention, also emerging with a lot of good works in recent years whose designers are from design strength of oversea or Shanghai, Beijing, Shenzhen and other places, but not local. This shows that the development of local design strength benefits from policy promotion, but the growth of local design strength can not be driven only by policy promotion and local design strength needs self-learning and profound consideration. Chongqing of Southwest China where more than a dozen of local five-star hotels are built up every year has become the battlefield for various design strengths during recent years, which indicates how hot its local design market is, but such a popular design market fail to breed symmetrical design works. The reason is worthy of our profound consideration. Comparing with common view over the impact by Hong Kong design on the Mainland's design, in particular on Shenzhen design, we also focus much on impact by Taiwan design strength on the Mainland's design, but we believe our concern is still not enough. Taiwan design, not only impacted by traditional Chinese culture but also very international, in my opinion, is the sui generis one and only way of Chinese contemporary design. Therefore, in this book, we particularly set up one chapter for Taiwan design so as to highlight its importance, and also hope to attract the further understanding of Taiwan design industry.

Design layout!

目录 A

002 序 INTRODUCTION

　　设计版图！ /Design layout!

010 华北 REGION NORTH CHINA

012 北京东方爱婴 /Eastern AiYing Preschool Center

018 北京波纹 /RIPPLES in Beijing

026 瑞居 /Gallery Hotel

034 光耀威海文登度假村北京售楼处 /
　　Guang Group Weihai Wendeng Resort, Beijing Selling Office

040 Groupm 群邑北京办公室 /Groupm Beijing Office

046 经典国际设计机构（亚洲）有限公司办公室 /
　　Office area of Classic International Design Agency (Asia) Co., Ltd

054 腾讯科技（第三极）办公楼 /
　　Tencent Technology (the third pole) Office Building

058 天津国贸销售中心 /Tianjin International Trade Sales Center

064 天津金泰丽湾嘉园 56# 单体售楼处 /
　　Interior Design of Sales Centre #56 of Tianjin Ecological Bay

072 天津津澜阙售楼处 /Sales Office of Jinlan Towers in Tianjin City

080 茗汤温泉度假村 /MING-TANG Hot Spring Resort

088 凰茶会 /Phoenix Tea House

094 华中 REGION CENTRAL CHINA

096 郑煤仁记体检院 /Renjihealth Checkup Hospital

102 武汉别墅样板房（B1户型）/Molding Show House

108 武汉武商摩尔城电影院 /Wuhan Pixel Box Cinema

116 武汉畅响会所 /Wuhan Echo Club

124 隐庐 /Hidden Cottage

132 梦网景园 /Dreamy Garden

136 华南 REGION SOUTH CHINA

138 光耀候鸟高尔夫球场中式别墅 /
　　Chinese Style Villa in Guang Group Bird Gold Course

144 深圳 La Vie Pub /Shenzhen La Vie Pub

150 西帷办公室 /INCANA Office

156 易菲展馆 /YIFINI Exhitbition

162 雅士阁美伦酒店 /Yashige Elegant Hotel

168 与自然共生 /Living In Nature

174 招商金山谷工法展示厅 /

Merchants Golden Valley Engineering Exhibition Hall

180 东北 REGION NORTHESAT CHINA

182 吉林艺术学院艺术咖啡厅 / Art Café of Jilin College of the Arts

186 亚布力 Club Med 冰雪度假村 / Club Med Yabuli

194 西北 REGION NORTHWEST CHINA

196 马柯艺术家餐厅 / Make Artist Restaurant

204 西南 REGION SOUTHWEST CHINA

206 拉萨瑞吉度假酒店 / St. Regis Lhasa Resort

214 成都岷山饭店 / Chengdu Minshan Hotel

222 华东 REGION EASTERN CHINA

224 唐乾明月福州接待会所

   Tangqian Moon Fuzhou Reception Club

232 熹茗茶叶会所湖东店 / Ximing Tea Club Hudong Shop

238 一信(福建)投资办公室 / Yixin (Fujian) Investment Office

244 静茶西湖店 / Taste of Tea and Buddha

248 天瑞酒庄 / Tianrui Winery

254 厦门光合作用 / Xiamen Guanghezuoyong

260 天墅销售中心 / Tianshu Sales Center

268 YS工作室 / YS Studio

274 尊墅别墅 / Distinguished Villa

278 厦门海峡国际社区原石滩SPA会所 /

   Original Rocky Beach SPA Club of Xiamen Channel

   International Community

284 巴厘香墅黄府 / Bali Incense Villa Golden House

290 CASA BELLA 家居美学馆 / CASA BELLA Home Art Center

294 合肥卡伦比咖啡连锁 / Hefei Kalunbi Coffee Chains

302 济南普利售楼处 / Jinan Puli Sales Office

310 杭州西溪MOHO售楼处 / Hangzhou Xixi MOHO Sales Office

314 千岛湖洲际度假酒店 /

   InterContinental One ThousandIsland Lake Resort

322 荣轩茶社 / Rongxuan Teahouse

330 g\g jeans宁波城隍庙旗舰店 / g\g jeans brand men's clothes

338 滕头投资公司生态楼 /

   Tengtou Investment Company Ecology Building

002 序 INTRODUCTION

设计版图！ Design layout!

酒店 HOTEL

026 瑞居 Gallery Hotel

080 茗汤温泉度假村 MING-TANG Hot Spring Resort

162 雅士阁美伦酒店 Yashige Elegant Hotel

186 亚布力 Club Med 冰雪度假村 Club Med Yabuli

206 拉萨瑞吉度假酒店 St. Regis Lhasa Resort

214 成都岷山饭店 Chengdu Minshan Hotel

314 千岛湖洲际度假酒店
InterContinental One ThousandIsland Lake Resort

办公 OFFICE

040 Groupm 群邑北京办公室 Groupm Beijing Office

046 经典国际设计机构（亚洲）有限公司办公室
Office area of Classic International Design Agency (Asia) Co., Ltd

054 腾讯科技（第三极）办公楼
Tencent Technology (the third pole) Office Building

150 西帷办公室 INCANA Office

238 一信（福建）投资办公室 Yixin (Fujian) Investment Office

268 YS工作室 YS Studio

住宅 HOUSE

102 武汉别墅样板房（B1户型） Molding Show House

132 梦网景园 Dreamy Garden

138 光耀候鸟高尔夫球场中式别墅
Chinese Style Villa in Guang Group Bird Gold Course

168 与自然共生 Living In Nature

274 尊墅别墅 Distinguished Villa

284 巴厘香墅黄府 Bali Incense Villa Golden House

餐厅 RESTAURANT

124 隐庐 Hidden Cottage

196 马柯艺术家餐厅 Make Artist Restaurant

茶馆和咖啡厅 TEA & COFFEE SHOP

088 凰茶会 Phoenix Tea House

182 吉林艺术学院艺术咖啡厅 Art Café of Jilin College of the Arts

232 熹茗茶叶会所湖东店 Ximing Tea Club Hudong Shop

244 静茶西湖店 / Taste of Tea and Buddha	Tangqian Moon Fuzhou Reception Club
294 合肥卡伦比咖啡连锁 / Hefei Kalunbi Coffee Chains	260 天墅销售中心 / Tianshu Sales Center
322 荣轩茶社 / Rongxuan Teahouse	302 济南普利售楼处 / Jinan Puli Sales Office
	310 杭州西溪 MOHO 售楼处 / Hangzhou Xixi MOHO Sales Office
商业空间 BUSINESS SPACE	174 招商金山谷工法展示厅 / Merchants Golden Valley Engineering Exhibition Hall
018 北京波纹 / RIPPLES in Beijing	
156 易菲展馆 / YIFINI Exhitbition	
248 天瑞酒庄 / Tianrui Winery	**文教医疗空间** Culture and Education for Medica
254 厦门光合作用 / Xiamen Guanghezuoyong	012 北京东方爱婴 / Eastern AiYing Preschool Center
290 CASA BELLA 家居美学馆 / CASA BELLA Home Art Center	096 郑煤仁记体检院 / Renjihealth Checkup Hospital
330 gxg jeans 宁波城隍庙旗舰店 / gxg jeans brand men's clothes	108 武汉武商摩尔城电影院 / Wuhan Pixel Box Cinema
售楼处 SALES CENTER	**酒吧和会所** BAR & CLUB
034 光耀威海文登度假村北京售楼处 / Guang Group Weihai Wendeng Resort, Beijing Selling Office	116 武汉畅响会所 / Wuhan Echo Club
058 天津国贸销售中心 / Tianjin International Trade Sales Center	144 深圳 La Vie Pub / Shenzhen La Vie Pub
064 天津金泰丽湾嘉园 56 单体售楼处 / Interior Design of Sales Centre 56 of Tianjin Ecological Bay	278 厦门海峡国际社区原石滩 SPA 会所 / Original Rocky Beach SPA Club of Xiamen Channel International Community
072 天津津澜阙售楼处 / Sales Office of Jinlan Towers in Tianjin City	338 滕头投资公司生态楼 / Tengtou Investment Company Ecology Building
224 唐乾明月福州接待会所	

REGION NORTH CHINA

华北

北京东方爱婴

项目地点：中国北京市
设计单位：SAKO建筑设计工社
照明设计单位：Masahide Kakudate Lighting Architect & Associates,Inc.
设计师：迫庆一郎、藤田悠
建筑面积：343 m²
主要材料：壁纸、MDF板、木地板、PVC
项目时间：2010.08 – 2010.12
摄影师：Misae HIROMATSU - 锐景Photo

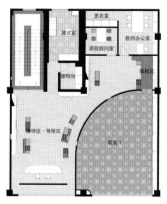
2F PLAN
S=1/150

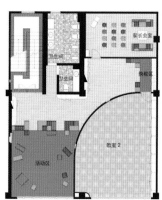
3F PLAN
S=1/150

东方爱婴早教中心是以0-3岁的宝宝为对象的早期教育设施，从安全要素考虑，将一切角全部倒成圆角。将长方形的角倒成圆角的"FILLET"用于家具、墙壁，又将其图案和地儿反转表现出来的"STAR"这一生动的图案印刷在地板上，在地面上铺展开来。各教室使用不同色系的颜色渐变的色调，创造出童趣活泼的气氛。

Eastern AiYing Preschool Center

Eastern AiYing Preschool Center is an organization for early education of children from 0 to 3 years old. In consideration with safety, all angles in the Center are converted into round ones. The "FILLET" to convert rectangular angles into round ones is applied on furniture and wall. The figure on it is reversed to become a lively "STAR" which is printed on the floor and spread out.

Each room applies gradually changing hue in different color systems, to create an interesting and lively atmosphere.

北京波纹

项目地点：北京市
设计单位：SAKO建筑设计工社
照明设计单位：Masahide Kakudate Lighting Architect & Associates, Inc.
设计师：迫庆一郎、角仓弘明
建筑面积：1960 m²
主要材料：MDF板、亚克力板
摄影师：Misae HIROMATSU - 锐景 Photo

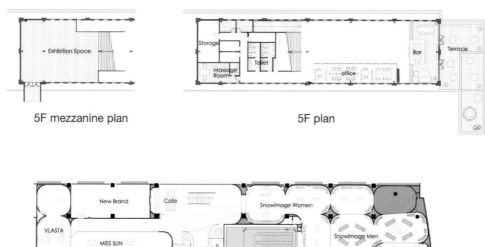

这是一座服装公司面向客户的展厅。以"波纹"为主题，将7个品牌向世界发布的样子在高14m宽60m的室内立面上表现出来。因为是冬装的品牌所以以"冬"和"雪"为主题，让1楼多个品牌展厅具有统一感。2楼是公司办公空间，多功能厅（用于服装发布会，展览）和BAR等配置在像纽约的LOFT那样的大空间中。6m高位置的梁裸露，使空间保持开放感，在墙壁上使用东西细长的平面，一侧是白色，另一侧设置汇集了PANTONE色彩的三角形波形装饰。从西侧看中二层，纯白的空间给人以娴静、端庄的印象，从另一侧看，则可感受服装品牌公司独特且色彩丰富的空间。

华北 REGION NORTH CHINA | 019

RIPPLES in Beijing

This is the clothes company's exhibition hall facing to customers. Taking "wave" as the theme, the seven brands are showing with 14 meters high and 60 meters wide in a way of releasing to the world. Due to the winter brand taking "winter" and "snow" as the theme, they have made the several brands in the first floor be unified. At the second floor, there are offices, multimedia halls (clothes release meeting, exhibition) and BAR and other equipment, which looks like the huge space of LOFT in New York. With 6 meters high exposed beams, the space has maintained an open style. There is adopted long and narrow plain on the wall. On the one side, it is white color, on the other hand, it is set much triangular waveform that collecting with PANTONE color. Viewing the second floor from the west side, the pure white gives people a quiet and elegant sense. From the other side to view the clothing brand's company are possessing colorful space.

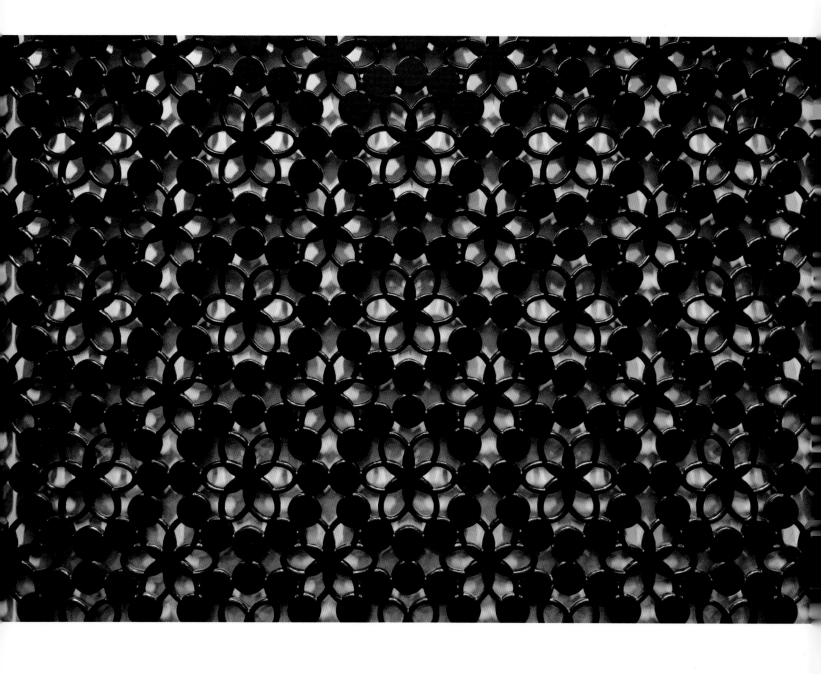

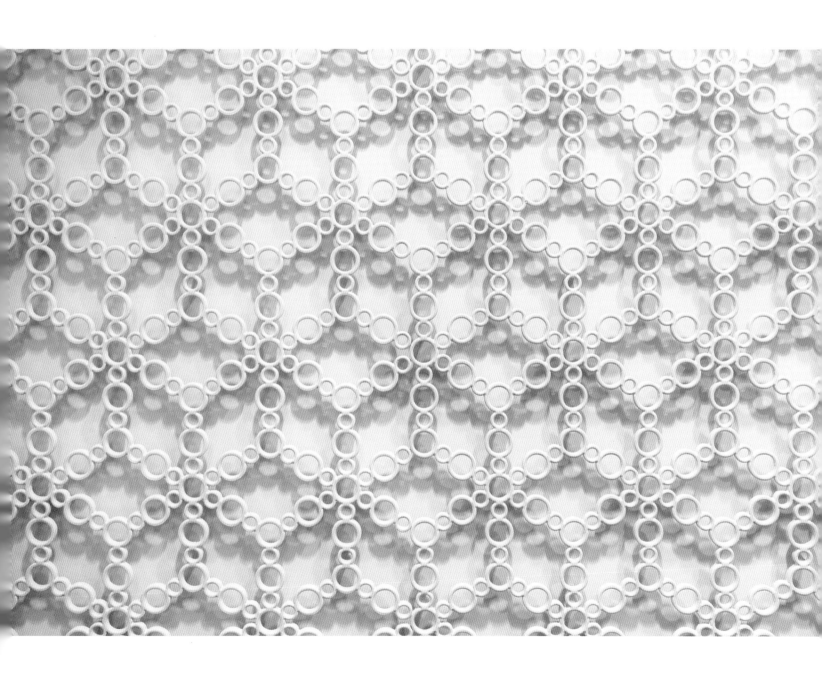

瑞居

　　瑞居是北京最具时尚、艺术气息的主题精品酒店之一。酒店特聘德国著名设计师精心打造，建筑外墙采用时尚的玫红色玻璃层叠，富有浓郁的当代设计及艺术时尚氛围，低调中尽显奢华。

　　瑞居的名称灵感源自明净祥和、吉祥福瑞、时尚靓丽、舒适宜居。酒店坐落于北京市朝阳区工人体育场院内——毗邻三里屯使馆区及CBD中央商务区。酒店备有40间高贵典雅的豪华客房及套房，所有客房均配有2台55寸液晶电视、高速无线网络、双线电话、iPod播放器、保险箱，客人可体会到品位极致的入住乐趣。瑞居定位为国际知名豪华酒店和时尚与尊贵设计。这里不但是北京唯一一家以典藏艺术精品为主题格调的艺术馆级高端酒店，更是京城独一无二的体验式艺术殿堂。

　　瑞居以尊贵、时尚及舒适为概念，强调乍然随意的含蓄风韵和不露声色的低调奢华。让人感觉恰如入住在现代艺术画廊之中。酒店设计的一系列灵感源自中国国剧京剧青衣脸谱，并巧妙融合了抽象化的传统中国细节处理手法于现代设计风尚之中。

项目地点：北京工人体育场
管理单位：红画坊（北京）酒店管理有限公司
设计单位：四川兴泰来装饰工程有限公司
设计师：Ricky
建筑面积：11 000 m²
主要材料：大理石、木地板、古铜色不锈钢

Gallery Hotel

Gallery Hotel is one of the most modern and fashion boutique hotel in Beijing. This hotel invites the famous designer of German to design. The wall of the building adopts fashion rose color glasses, full of modern design and fashion sense, in low profile but luxury.

The name of Gallery Hotel comes from bright and clear, lucky and happiness, fashion and charming, comfortable and livable. This hotel is located at Worker's Stadium, Chaoyang District, Beijing—close to Sanlitun embassy district and Central Business District. This hotel possesses 40 elegant and gorgeous suits. Each of suit s equipped with 2 sets of 55 inches LED TV, high speed wireless internet, Two-line phone, iPod player, strong box. Guest can enjoy the ultimate comfortable here. Gallery Hotel is positioned as an international well known luxury hotel and the sign of fashion and distinguished design. It is not only the exclusive art-like hotel taking boutique arts as the main theme, but also a unique experience type art gallery.

Gallery Hotel takes distinguish, fashion and comfortable as the concept, emphasizing on the naturally charm and unemotional luxury. You will feel live in an art palace. The inspiration of this hotel design comes from Chinese National Opera Tsing Yi face. It has merged the abstract traditional Chinese treatment ways into the modern fashion design.

光耀威海文登度假村
北京售楼处

项目地点：北京三里屯 SOHU
设计单位：立和空间设计事务所
设计师：贾立
参与设计：张韬、苗壮、高璇
建筑面积：218 m²
摄影师：高寒

蔚蓝的大海，金色的沙滩，圆润的卵石，蜿蜒的海岸，舒适的心情悠然而生。

碧绿的草地，繁茂的树木，散布的球洞，起伏的场地，惬意的感受不言自明。

不同的景致，相同的心情，构成了本案空间设计的灵感来源。

入口处，一排门洞型木纹铝的方通格栅组成的走廊将室内空间分隔开来，高尔夫体验区和接待区自然地散落在其两侧，走廊的尽头是整个参观体验的终点，也是洽谈区的所在。方格栅形成的高低起伏、大小结合的空间形式既打破了原有沉闷且毫无遮蔽的状态，又起到了划分功能的作用。走廊是整个项目信息展示的核心区域，项目沙盘展示和区域沙盘展示位于中心位置，形成整个空间的视觉中心。单体沙盘和静态展示盘有规律地排列在走廊两侧，丰富了视觉体验，也满足了销售的讲解动线。进入高尔夫体验区，高尔夫球和球杆看似无规律地悬挂在展示区域内，传递出休闲轻松的气氛。卵石型的门框和窗框则与家具造型产生了良好的互动，相得益彰。

整个售楼处在色彩运用上力图打破狭长空间带给人的压迫感，柔和的浅色系成为了首选。竖向排列的海蓝色背漆玻璃成为整个空间的底色，将室内各个元素映射其中。蓝色延伸空间，又让人们感受到海的平静与深邃。卵石状异型灰白、灰绿色沙发交替排列，与透明亚克力坐凳搭配，虚实结合，层次分明。色彩丰富的海洋生物雕塑点缀其中，生机勃勃，趣味丛生，让人浮想联翩。

Guang Group Weihai Wendeng Resort, Beijing Selling Office

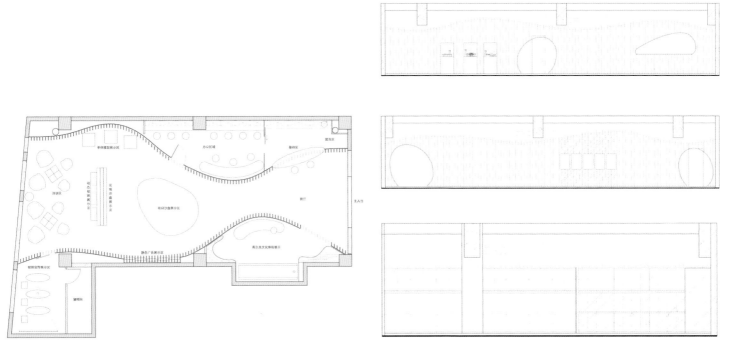

Blue sea, golden sand, round pebbles, winding coastline, make you have a comfortable mood.

Green grassland, flourish trees, scattering holes, the ups and downs ground, great happiness will fly into your heart.

Different landscape, same mood, they have consisted the inspiration of this space design.

At the entrance, a line of hole-type wooden texture aluminum grid corridor has separate the indoor space. The golf experience area and the reception area are scattering on both side naturally. At the end of the corridor, there is the terminal of the whole experience and also the negotiation place. The ups and downs formed by grid has broke the traditional boring open situation. It also plays an important role on dividing the functions. The corridor is the core area of showing this entire project information. The project display sand table and area sand display table are located at the center place, forming the visual center of the whole space. The single sand table and the static display table are listed along the both sides of corridor, which has enriched the visual experience and satisfied the selling introduction. Enter the golf experience area, the golf ball and the golf clubs are hanging there naturally, showing a leisure and relaxation mood. The pebble-like door and windows has generated a good interactive with the furnitures.

The entire selling office has broke the boring sense of the long and narrow space in color. The soft light color are put first. The vertical sea blue painting glass has formed the basic tone of this space, merging all kinds of elements into a single whole. The blue color has extended the space and let people feel the peace and deep of sea. The pebble-like gray and white color as well as the gray-green color sofa are listed there. Coupled with the acrylic chairs, they are looking colorful and orderly. The colorful sea lives sculptures are dotted the space, making the space vibrant and full of interesting.

Groupm 群邑
北京办公室

项目地点：北京金宝大厦
设计单位：Mi2
设 计 师：陈宪淳
建筑面积：1200 m²
主要材料：灰镜、单反玻璃、钢化玻璃、水切割铝板、穿孔喷绘、宇宙灰石材
设计时间：2011年
摄影师：孙翔宇

Groupm 从属 WPP 集团，下辖 Mediacom、Mindshare、MAXUS、MEC 等国际知名的 4A 广告传媒公司。这是北京的办公室，简单流畅的现代开放办公风格，巧妙利用全新的工艺赋予灰镜、铝板、石膏板、彩色喷绘等传统材料全新的生命力，空间结构实用至上，好用然后好看。这就是商业办公空间最永恒的出发点。

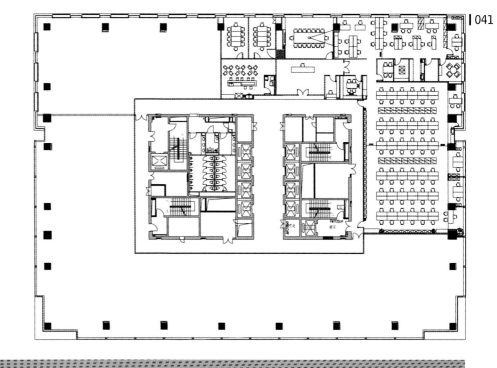

Groupm Beijing Office

Groupm is subjected to WPP Group, affiliated to Mediacom, Mindshare, MAXUS, MEC and other international well-known 4A advertising media company. This is Beijing Office. Simple and smooth modern open style office, the designer has used the new technology process on the gray mirror, aluminum, Gypsum board, color injet and other traditional materials with new vitality. The spatial structure puts practical first and then focuses on the good-looking. This is the enternal strarting point of the commercial office space.

经典国际设计机构（亚洲）有限公司办公室

"慢"是"快"的基础，只有习惯"慢生活"，才能够快速准确找到定位，而不会迷失自己。要慢下来（Slow），是因为"快"让人错失了很多美好的事物。所以我们倡导：慢生活、慢餐饮、慢睡眠、慢工作、慢情爱、慢社交、慢读书、慢运动、慢音乐、慢设计……

我们秉承这一理念，将"慢"的理念延续到我们的办公空间。首先是选址，北京的空气净化器——森林公园成为我们的最佳选择。这栋建筑位于森林公园的腹地，依山傍水，独立清幽，完全隔绝都市的繁杂和喧闹，是真正意义上的室外桃源。

我们尽量尊重原有建筑的空间结构，错层、高达6 m的空间高度、三角形的采光顶，都得以保留，原有的不规则结构梁成为我们的照明基座。狭长的残疾人坡道成为材料区和文印中心。而宽敞的户外露台成为绝佳的休闲和放松的区域。

我们使用最单纯的设计语言，把安静的气氛融入空间之中，置身室内，窗外的自然美景是最大的视觉重心。我们可以静观微风吹过，枝叶飘摇。水光天色，山重树茂，无不快哉。室内家具和艺术品的选择也同样遵循"慢设计"的理念，只有被称之为经典的才能称为空间的主人，明式圈椅、The Chair、Y Chair、Ghost Armchair轮番登场，Pop Art、北魏造像、当代艺术交相辉映，共同谱写一组和谐的乐章。

置身这样的空间之中，心会自然地安静下来，快的节奏和习惯会慢慢远去，我们会更清晰地思考，更深入地研究，以致更精准地处理设计中的所有关系，努力创造更具深度的作品。为中国设计走向世界贡献自己的微薄之力。

项目地点：北京市朝阳区
设计单位：经典国际设计机构（亚洲）有限公司
设计师：王砚晨、李向宁
建筑面积：800 m²
主要材料：白色乳胶漆、透光膜、钢板、白色人造石、玻璃、白色铝塑板、自流平地面

Office area of Classic International Design Agency (Asia) Co., Ltd

"Slow" is the basic of "fast". Only used to "slow life", can you be able to find your position quickly and accurately without losing yourself. We shall slow our steps, because "fast" let us miss a lot of beautiful things. Therefore, we call on slow life, slow food and beverage, slow sleep, slow work, slow love, slow communication, slow reading, slow sports, slow music and slow design.

Upholding this concept, we put "slow" into our office area. The first thing is to choose address. Beijing's air purifier—Forest Park is our best choice. This building is located at the center of the Forest Park, surrounding by mountains and waters, independent and quiet, which has isolated from the bustle and hustle city. It is the true outdoor place for people.

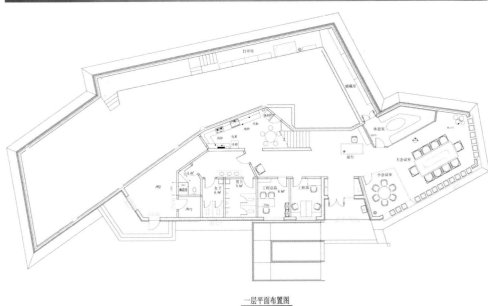

一层平面布置图

We try our best to respect the spatial structure of the original building. The overlapped, 6 meters high, triangular skylights are all preserved. The irregular structure beam has formed our light base. The long and narrow disabilities road becomes the material and cultural printed center. However, the outdoor terrace has become the perfect leisure and relaxation areas.

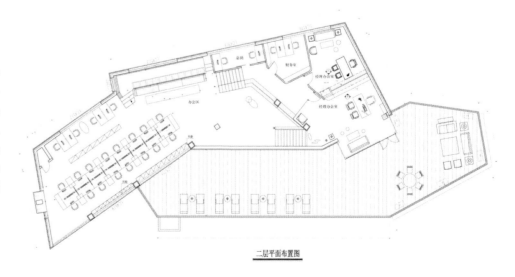

二层平面布置图

We are using the most simple and purity design language to put the quiet into the space. Standing in the room, the nice landscape of outdoor is our biggest visual center. We can enjoy the gentle breeze that makes leaves swaying. Water and sky, mountain and trees, how joyful it is! The interior furniture and art products are all in line with the "slow design" concept. Only the classical can be called the owner of the space. The chair, Y Chair, Ghost Armchair are walking on the stage one by one. Pop Art, Beiwei Dynasty statues, contemporary art are interacted with each other, showing a harmonious piece of music.

Standing among such space, our heart will be calm down slowing. The fast speed life and habit will leave us gradually. We will have a clearer meditation and deeper research to handle all the relationship of the design to create a better works.

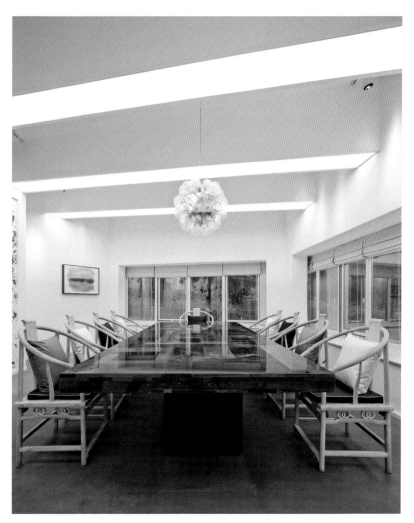

项目地点：北京第三极大厦7层-11层
设计单位：北京艾迪尔建筑装饰工程有限公司
设计师：罗劲
参与设计：张晓亮、黄丽元、张清、董欣亮
建筑面积：**21330 m²**
主要材料：自流平地面、石膏板、钢化玻璃
摄影师：高寒

腾讯科技（第三极）办公楼

腾讯科技一直信任艾迪尔能够实现他们对办公空间的设计需求——活泼、和谐、创新、追求卓越。本次北京第三极办公室的项目设计主题是"最美的宇宙奇观",关键词是"神秘、无限、美丽"。我们分别用"银河"、"陨石"、"星座"、"彗星"、"流星雨"、"极光"、"星云"的具体概念塑造了"神秘的"各层办公空间。色彩上,我们不但延续使用了以往项目中常用的腾讯logo色系中艳丽的黄色和橙色,还增加使用蓝色、紫色来表现宇宙主题。整体风格现代又不失稳重,活跃又不失雅致。

Tencent Technology (the third pole) Office Building

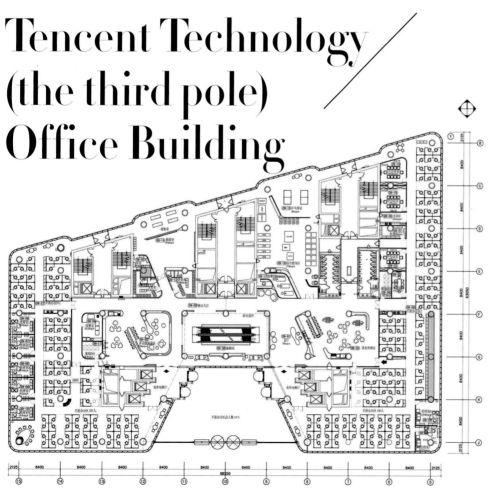

Tencent Technology has been trust IDEAL to achieve their design requirements for office space--lively, harmony, innovation, pursuit of excellent. The third class office building project in Beijing takes design theme of "the most beautiful universe spectacle". The key words are "mysterious, infinite and beautiful". We adopt "galaxy", "meteorite", "constellation", "comets", "meteor rain", "aurora", "nebula" to shape the "myterious" office space. In terms of color, we have continued the bright yellow and orange color used by previous projects of Tencent Technology logo. In addition, we have added blue, purple color to show the universe theme. The whole style is modern and calm, active and elegant.

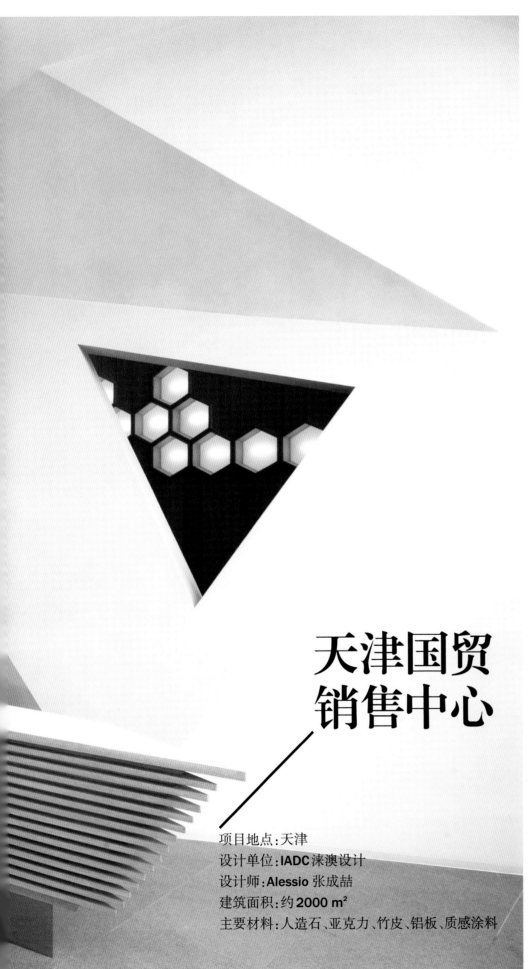

天津国贸销售中心

项目地点：天津
设计单位：IADC涞澳设计
设计师：Alessio 张成喆
建筑面积：约 2000 m²
主要材料：人造石、亚克力、竹皮、铝板、质感涂料

灵感来源于中国古代山水画的意境，山水间云雾缠绕，远处森林、鸟鸣，平静的水面，一叶叶扁舟，怡然自得，构成了和谐的画面和生活境象。木的背景，几片喻意山峦起伏的构造物，吧台象一艘艘木舟，高低错落，组成了空间的基本形态，抽象的演绎，简洁而富于张力的设计语言，更像是一幅空间的水墨画。

建筑由二个楼面组成，因空间比较狭长和不规则，在初期构思中，以比较单纯的三片构造物分别构成了空间的主要功能，多媒体室，接待区和酒吧区，洽谈区。而在立面上的白色构造物使狭长的空间充满了韵律感，视觉上更富有张力，也成为空间的基本构成元素。而背景则以素净的竹皮连贯了整个空间，连续的"蜂巢"造型喻意"建造栖居"，与白色的空间一起共同营造了和谐、温暖的氛围。连接一、二层楼梯间的金属构造同样延续了蜂巢的主题，金属与发光的造型带有未来主义色彩。单纯的材质和简约的造型烘托主题，有机形态富于动感和趣味，为空间带来变化，线条简洁的家具和灯饰也为空间增添了优雅的品味与现代感。

Tianjin International Trade Sales Center

Inspired by the ancient Chinese landscape painting, the beautiful mountain and clear water are winded by clouds and fog. The distant forests, birds, quiet waters, a leaf of boat have constituted a harmonious picture and living environment naturally. The wooden background, a few slices that symbolized mountains, the wood-boat-like bar, they are palced high and low and formed the basic shape of the space. Abstract interpretation, concise and powerful design language, it more likes a Chinese brush drawing of space.

The building is constitued with two floors. As the space is narrow and irregular, at the initial consideration, the designer takes three things to separate the main function area of this space, such as media room, reception area, bar, and negotiation area. While, the white long material has mede the narrow space full of rhythm, and powerful in visual. They are also become the basic elements of the space. While, the background takes the bamboo trougbout the space. The continuous "honeycomb" means "build and live". Cooperating with the white color space, they have formed harmonious and warm atmosphere. The metal structure that connects the first and second floor continued the theme. The metal and the glowing furnitures has a sense of futurism.Taking the simple materials and simple shape to reflect the theme, the organic form seems full of dynamic and interesting, which has brought many changes to the space. The clean lines furniture and lighting has added elegant and modern sense to this space.

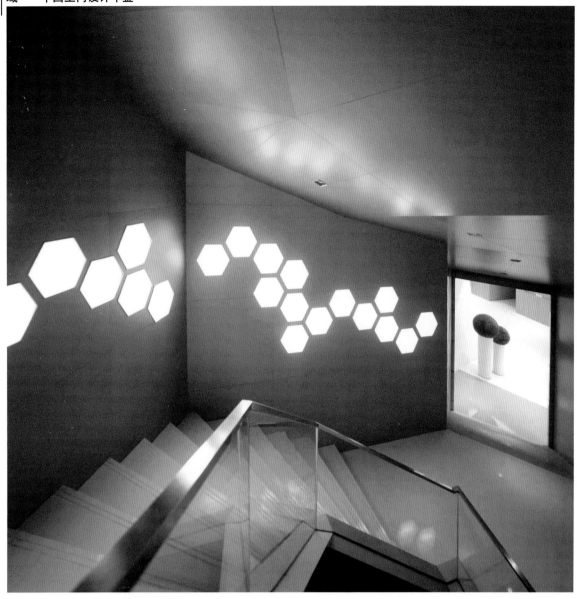

求,并建立与环境的联系。

"艺术家建设自己的小天地,而建筑师创造人们共享与包罗万象的世界。艺术家总是暗示,而建筑师经常本能地想象,那些能让人类与自然和宇宙和谐共处的空间。这就是点燃建筑师激情的燃料。建筑与环境的和谐增强了我们的认知能力,让我们能够全情投入。是建筑将我们从公共生活带入了私密(属于自我)的世界。"建筑师Davide Macullo如是说。

对于本案设计者来说,一切设计的最终目标也就是让人们得到幸福。为了实现项目而进行的每一个创造性的、智慧的和实际的行动都决定着它是否能为生活带来快乐。在这一层面上,我们看到了这一项目带给生活的真正品质。

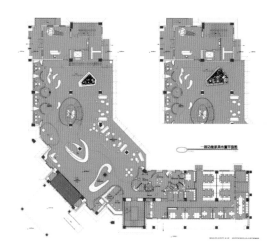

Interior Design of Sales Centre #56 of Tianjin Ecological Bay

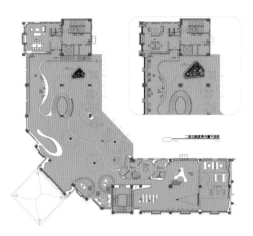

Jointly designed and completed by Davide Macullo, the Swiss architect, and Beijing Yidi Building Design Firm, the sales office of Tianjin Ecological Bay contributes to easy, pleasant spatial atmosphere; as well as accurate professional quality at the same time. Full of a home-like sense, this space is in both internationalization and urbanization at the same time, warmth and novelty, and focused on functionality and experience. It is flowing, impresses one's feelings, sprawling, clear, hospitable, intimate and emotive.

First of all, the sales office shows its formal and typical side and helps to convey hospitable intentions to the outside world; therefore, also indicates an excellent opportunity of exchange and relationship establishment between people. This reproduced theme has been converted into the dynamic geometric figures which is captured, concentrated and extracted out by richly expressive and rigorous architectures. Therefore, the geometric figures are organically flowing over the interior and exterior spaces; and their intervals are separated by the transitional space. The sequence combinations of different spaces, from small to big, near to far, private to open, through the internal volume and forms, result in continuous inspiration and impression, where people unknowingly get the most enjoyable or pleasant feelings. Further more, space setting and allocation are considered for flexibility of usage in order to expand its scope of use (after the end of sales, it will be continuously used as Art Museum and Sino-Swiss cultural exchanges center), thus leads to new creation and experiences. In this sense, the space is not rigid but continuous, which shall be constantly stimulated by new discovery by the users, and then brew the new or even unknown content.

The natural form of space generated from the urban texture is characterized by two floors connected through flow curves to create a continuous interface and thus results in continuous dynamic spatial lines, where people's mood to stroll idly in courtyard and elegant posture are revealed while climbing up the stairs, and they feel more and more close to the space step by step. Here, people will feel to be surrounded and continuous changes of spatial scale, as if they travelled to a new place appears with ceaseless surprises; it allows us to become more sensitive to the passage of time. Stay in a place to stimulate the senses for a long time helps people to thus enlarge their concept of time. Overlooking the far scenery from the second floor, one will see the outdoor garden at the same time; overlooking outward from the negotiating room, one will find a large clean area extending toward one direction, and then look back to the garden, once again see the sandbox area. This kind of changeable view, even within a small space, can also deepen people's sense of space, whose life thus becomes a trip of time and senses.

In a sort of sense, the spatial function here is about people themselves – makes them feel to be embraced and surrounded. Designer has expressed space and environment through overall trend of taxis, firmness, sparse density and brightness, etc under space-dimension arrangement, as well as overall contact between space and people. In the internal structure of the building, the separated space is organically organized between these divisions, from interval of those levels, we can feel fore-and-aft infiltration, transparent overlap from top and bottom, oblique connections as well as light from different directions, so that from any location, we can feel views nearby, far away or in

middle distance at the same time. Furthermore, the external space of sales office is also used as internal part, built on three-dimensional connections with space relating to both scenery orientation and light. Thus, the people' spatial experience is rooted in a large environment.

Space should become people's sensorial extension, as if our particularly keen sense of smell, hearing and vision to perceive three dimensions, a huge mouth and the longest arms to affectionately embrace, and as long as the organism to help our brain operation is alive, we will become wise. The project really makes us aware that the space is the projection and reflection of feelings, the perspective alienation and the place where we are eager to follow our heart and be ourselves.

Resulting from the meticulous work of the designer step by step, the space has a sculptural tactility so as to create an environment for people to immerse all their feelings. However, this sense of sculpture is not made deliberately; but deriving from designer's interests in and attitudes to its internal aeriality so as to be converted to observation way at deeper level, analyze people's needs in aspect of density, and establish linkages with environment.

"Artists construct their own small world, while architects create the coverall world to share with people. Artists always imply, while architects often instinctively imagine those spaces for people in harmony with nature and universe. This is the fuel to light architects' passion. The harmony between architecture and environment helps to enhance our cognitive abilities and enables us to be whole-hearted. Architectures bring us from public life into private (belonging to ourselves) world. ", architect Davide Macullo said.

For the designer of this project, the ultimate goal of all designs is to make people happy. Each creative, wisdom and practical action in order to achieve the project determines whether it can bring happiness to people's life. At this level, we see the true quality brought by this project to life.

天津津澜阙售楼处

项目地点：天津海河河畔
设计单位：北京空间易想艺术设计有限公司
建筑面积：**1160 m²**
主要材料：石材、亚克力、铝板、皮革

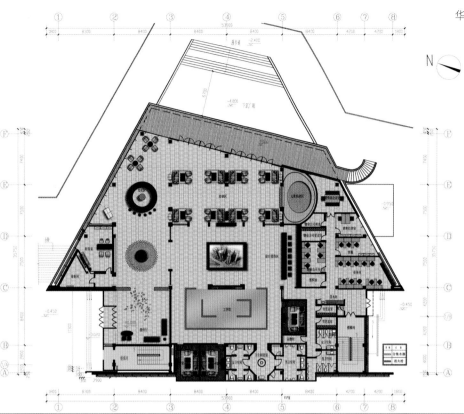

本案位于天津海河河畔，具有优越的自然景观，项目定位为品牌高端产品。为了打造高品质的奢华空间，我们对原有建筑及周边环境的分析，尽力将售楼处周边的景色与室内的环境相结合，使之成为有机结合的整体。原建筑为天津成立的首个游艇俱乐部"天津湾游艇会"所有，地处天津湾水上公园。利用自然河滩坡地和周边绿化，使建筑规划融于河岸地形中，形成独特的绿色景观效果。凭栏临风，海河上水鸟鸣叫，扬起片片浪花，偶尔还能看到出海兜风的游艇迅速滑过水面，留下叠叠雪白的浪花。为了保留这种特质，我们在空间上塑造了开放而通透的空间气氛，将室外的环境引入室内，使室内外环境完美地结合在一起，以低调奢华的气质凌驾于海河湾畔。

整个售楼处追求的不仅是纯功能性，更极大地挖掘其视觉效果，冷与暖的对比、虚与实的搭配、动与静的转换，都相处得非常融洽。

在色调上以舒适的灰白色为中间调，局部配以深色来点亮空间，使整个空间华丽而沉稳。灰色的木纹石纹理清晰，经过重新排列，形成了一幅天然图案，而现代的家具配饰和雕塑感很强的墙面都相处得非常融洽。

材质上整个内部采用简洁的灰色石材，局部配以深色、具有线条感的装饰屏风，形成繁简对比，构成了一幅现代抽象画。香槟金色的皮质软包，使整个墙面在简练之中透显着严谨精致的细节、使空间显现出一种低调的奢华。

接待前厅墙面采用一定厚度的亚克力柱

墙，形成了具有围合感的纯净空间，塑造出若隐若现的层次感。

模型区使用纵横交错的螺纹钢，形成一扇扇奇妙的现代屏风，把原本的不规则空间围合出具有建筑美感的秩序空间。

整个洽谈区天花采用几何形铝板排列成淡淡起伏的形态，犹如海浪般在天花翻滚，又犹如从水吧台中生长的一束烟花，布满整个洽谈区，使整个空间动中有静，静中有动，在简单中不失现代韵味。

工学馆外皮镂空的装饰墙与灯光一起构成了一个现代并富于光影和雕塑感的空间。内部整体风格为现代风格，色调以灰黑色为主，利用特殊的灯光效果，营造出一个现代、充满科技感的展示空间，为行走其中的观看者不断呈现一个具有现代艺术感的空间。展示区的整体色调为暗灰调子。展区光源主要以暗藏的线形光源为主，有效地避免了眩光，并能充分烘托展区的空间气氛。展台区部分采用了局部照明，采用本色白光，接近自然光的颜色，可以保证展品效果不会失真。

在整体配饰设计中，为了契合项目的特质与地理特点，我们选择了海洋主题，利用现代的材质及手法来表现。波澜起伏的天花中悬挂着银色金属鱼群，用抽象的形态表达了欢快自由的主题。在饰品的选择上也采用了金属质感的抽象雕塑，表现出太湖石、漂流木、水草、船帆等主题。天津市是有多年历史的港口城市，在设计之初我们就希望能在船厂寻找一个有悠久历史的旧船锚做为门口主要雕塑，最终并未找到，这也是此项目中留有的一点遗憾了。

Sales Office of Jinlan Towers in Tianjin City

This case is located at side of Haihe River of Tianjin, boarding excellent natural landscape. This project is positioned as brand high-level product. In order to create a high-quality luxury space, we have analyzed the original buildings and the surroundings to make them keep same tone with the internal environment. The former building is owned by "Tianjin Yacht Club", which is first yacht club of Tianjin, located in Tianjin Beach water park. Utilizing the nature river beach and the surrounding greenland, they have put the construction into the beach design to form a unique green landscape. Leanning on a railing, the birds are singing on the Haihe River, raising patches of waves. You can see the yacht drive over the surface quickly, leaving piles of white spray. In order to keep this characteristic, we has shaped a open and transparent space to introduce the exteranl environment into the door. Combined the internal and external environment together, they are showing the Haihe River beauty in a low-profile but luxury way.

The whole sales office is pursuiting not only the pure functionality but to tap its visual effects more. The contrast of cold and warm, virtual and reality, dynamic and static, are all get along with each other very well.

It is adopted the gray and white color as the midtones. Partial parts are coupled with dark color to light the space, letting the whole space luxury and calm. The gray wood texture and stone texture are clear. Through re-arranged, they have formed a natural pattern. They are getting along with the modern furniture and sculptural.

It is adopted the gray stone as the materials. Partial parts are coupled with dark color decorative screen, forming a contrast of simple and complexity. It is a piece of modern abstract art. The champagne gold leather soft package has made the whole wall shown a rigorous and exquisite sense. What a beautiful and luxury the space it!

The wall of reception hall adopts a certain

thickness arcylic column wall, forming a enclosure space sense, and shaping a looming sense.

The model area adopts the intercross steel, forming pieces of modern screen.
They have made the irregular space be a aesthetic architectural space.

The entire negotiation area's ceiling is adopted geometric aluminum to form a wave-like rolling state, liking the water splashing in the ocean, or a bunch of fireworks grow from the water bar. They are fulfilled with the entire negotiation area, making the entire space moving in the quiet environment, simple but meaningful.

The hollow decorative walls of engineering building has formed a modern and shadowy space coupled with the lights. Its internal style is modern, taking gray and black as the tones. Utilizing the special light, they have create a modern and sentific space. It is a modern artistic space showing to the visitors. The exhibition lights are almost hidden, which has avoide the light glaring directly and can fully add fun to this space. The exhibition part adopts partial light. It takes the white color light, which is close to natural light and can guarantee that the exhibits effect without distortion.

In the overall accessories design, in order to in line with the project's characteristics and geographical features, we chose the ocean as the theme. Utilizing the modern materials and approaches to perform it. On the waving ceiling pattern, there is hanging groups of silver and golden abstract fish, which use abstract state to express the cheerful theme. On selecting of the decorated products, they are also adopting the metal sense abstract sculpture, showing eroded limestone, driftwood, plants, sails and so on. Tianjin is a city near the sea with long history. At the initial design, we hope that we can find a old anchor ship with long history as the main sculpture. But we did not find such ship ultimately, which is also a regret of this project.

茗汤温泉度假村

近年来由于大陆都市发展迅速成长，休闲旅游的需求量增多，观光产业成为积极开发的重点项目。借由与北京及天津大约一个小时车程距离及拥有丰富的温泉资源的优势，霸州被计划打造成温泉城，成为河北省环京津旅游圈内的重点项目。在整个温泉区内总共规划有6块开发用地被分配给6家投资团队，包括茗汤休闲度假集团。

如何借由环境带给人在心灵感受上的丰富性，而不是依靠气派及豪华，来重新定义五星级度假旅馆的质量是我们关心的议题及立场。在这个之上思考，塑造一个适合的环境是规划过程里首要的重点，然后是置入景观的主题与经验，最后才是建筑形式的考虑。对我们而言建筑最后的形式只是思考如何与周围形成的环境融合的一种结果。

基地大约18公顷（第一期有12公顷左右）。地形平整没有太大起伏，呈长方形状，冬季的寒冷季风会从西北方向吹向基地，而夏季的凉爽季风则会来自从东南方向。此外四季的气候变化分明所以景观很不一样。春天宜人、夏天炎热最高会达到30多度、秋天凉爽、冬季寒冷会下雪、最低会到零下10度；所以如何将度假泡汤与环境在四季变化里带来的经验结合是规划的重点之一。

在全区布局上，借由重新塑造的地形来应对气候对基地的影响：在北面有较高的地形如同屏障来阻挡冬季寒冷的西北季风，在南面则以低的地形引导夏季凉爽的东南季风进入基地。而不同的地形也形成了不同的景观类型及塑造了多样性的景观经验：丘顶—草原/山坡—树林/低地—温泉。最后环境形成之后，会馆及别墅在不同的地形置入形成不同景观主题的群落，例如温泉会馆，湖畔小屋或是林

Stone，石材

Wood，實木

HOTEL ELEVATION & MATERIAL，會所建築立面及材質

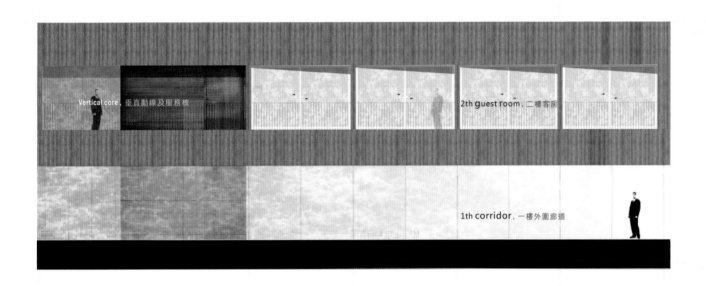

Vertical core，垂直動線及服務核
2th guest room，二樓客房
1th corridor，一樓外圍廊道

项目地点：中国河北省霸州市
设计单位：C T Design 城市设计
设计师：陈源盛、曾锦玲
参与团队：LPD Landscape Planning & Design, Japan、
　　　　　MICHIKO YOKOTA and LIGHI, Japan
主要材料：实木、石材、青石板、洗石子、卵石
项目时间：2010.11
摄影师：小雄梁彦影像有限公司

中木屋。

在景观布局上，水景在被丘陵围塑的低地形成之后，附近较高的丘坡成为林带，较低的坡地成为花带，湖畔较为平坦则成为草地。以温泉为主题的会馆被置入在三种不同的景观之间；水、竹林、花三种景观元素由外向内围塑空间。

在度假村里面，水是最主要的元素。从泡汤到SPA，从中庭内的水景到外围的景观湖及小溪，都是希望借由人与水在不同层面的接触去体验水的文化。水的系统是一个循环系统，二次利用的温泉水被加压至水景中庭的水墙及水瀑，由高往低、由内向外流动，由水景中庭流向外围的水池再往下到景观湖，最后进入小溪流到终点的蓄水池后再加压回流至内部水景中庭。

自然中的建筑、建筑中的自然，是我们会馆规划的主要概念。简而言之它并不是一个从强调造型及风格着手的建筑。我们希望塑造一个谦虚的能跟环境融合一致的空间，它的形式是不需要被强调及被注意到的。建筑形态的转换过程有3个重点。主要是希望找出能够与周围环境融合的最佳形态及与景观对话的最佳关系。

1. 如何打散巨大的量体形成线性的空间避免对环境造成冲击。
2. 如何运用线性的空间让空间与环境及景观的对话能够形成最佳关系。
3. 如何最后导成线性环状空间形成三度循环的最佳动线。

建筑有如一个线性的环状量体，置入到基地里面并不会将环境盖掉同时能带给空间与景观接近的最大机会。我们希望建筑空间本身就如同一个三度的循环动线，是一条能够将人带入及体验环境的廊道。温泉会馆的空间布局，基本规划为三个环状区域，主要是依据周围景观的属性来做配置。另外地面的高层也跟随周围地形的高度变化提升：北面大厅提高0.2 m，餐厅及泳池提高0.5 m，南面门厅提高0.8 m，客房提高1.1 m。

三个环状区域空间：

环一：大厅、Café、SPA及室内外泳池 - 水区

空间被水环绕，感觉像在水面之上。我们希望借由相临及延续的关系让室内的空间跟周围的水景融合；同时也借由温泉在冬季因温差产生雾气的现象塑造出让旅馆能够看起来就像是在水雾之中的感觉。

环二：餐厅 - 花区

餐厅紧邻花区；包厢区的外围是花坡，用餐区及包厢走道则围塑以樱花为主题的中庭。我们希望借由以花及干景为主题的景观来塑造餐厅的精致气氛。

环三：客房 - 林区

客房区在林带之中，空间被竹林围绕。选用竹子塑造林带除了是因为竹林的意境能跟空间的属性相符之外，主要是因为竹林能够形成高密度的自然保护带且具有耐寒的特性所以能够提供客房四季所需要的基本私密性，特别是针对1楼客房面对内部中庭的泡汤区所需要的隐密。

我们希望这个度假村是一个可以让人在视觉、听觉、嗅觉及触觉都能够感受的环境；就像是会听到水声、鸟声及风吹过竹林的声音，能感受到温泉及雾气，也能闻到花的味道。

MING-TANG
HOT SPRING RESORT

In recently, as the rapid growth of urban development from the mainland, and the improvement requirements of leisure and tourism industry, the tourism industry has become the major project being developed. As only one hour's drive from Beijing and Tianjin, as well as owning rich hot spring resources, Bazhou has been planned to built to a Hot Spring Town, becoming the key project of Beijign-Tianjin tourism circle in Hebei province. In the overall spring area, there are six parts for development, which have been assigned to six investment group, including Mingtang Leisure Resort Group.

How to bring wonderful feeling to peopla's heart by environment but not depend on the luxury and grand shape to re-define the quality of the 5-star hotel is our issues and position. Taking it as the basic, shaping a suitalbe environment is the key point of the planning process. And then, place the landscape's theme and experience. Finally is the coonsideration about construction type. For us, the construction type is only a result of merging with the surrounding environment.

The base is about 18 hectares (the first phase is about 12 hectares). The terraiin is plain without ups and downs. The cold winter monsoon blows from the northwest. While the summer sonsoon is from the southeast. In addition, the four seasons is clearly divided and have different landscape. Pleasant spring, hot summer which will has 30 degrees maximum, cool autumn and cold winter with snow and the minimum temperature will be lower than minus 10 degress. Therefore, how to make the resort connected with the changable four seasons is the key planning point.

On the overall layout, it is using the re-shape terrain to response to the impact of climate on the base. There is higher terrain as barriers to prevent cold winter northwest monsoon in the north. There is lower terrain to allow the cool summer southeast to enter the base. While different terrains have formed a different type of landscape and the

diversity of landscape experience:hill-grassland/hill slopes-woods/lowland-spa. Finally, after the environment is re-shaped, the club and villas will form different landscape theme community in different terrain, such as spas, lakeside cabin or forest huts.

In landscape layout, the water is placed at the lowland which is surrounded by hills. The higher slope is made into forest, and the lower slope is made into flower trip. The lakeside is more grassland. The pavillion taking spring as theme is place three different landscape. Water, bamboo forest, flower are the landscape elements from the internal to external.

In this resort, water is the most important element. From immersing in the hot spring to spa, from the court water landscape to the external landscape lakes and streams, we all hope people can experence the water culture in different aspects. The water system is to let the secondary hot spring water use for water wall and waterfall throught pressure. From high to low, from internal to external, waterscape flows to the external lake and finally enter the stream to the reservoir. Then the water will come back to the internal waterscape's circulatory system by pressure.

Construction in nature, nature in the construction, are the main concepts of our planning about the club. In short, it is not a construction only focusing on the style and shape. We hope to build a humble and harmonious space. Its shape is not need to be stressed and noticed. The architectural form has three key points during the conversion process. It is main hoped that we can find the best way to merge the environment and landscape.

How to break up the huge amount to form a liner space to avoid impact on the environment.

How to use the liner space to let the space and the landscape can form the best relationship.

How to form the liner space and an excellent 3-cycle moving lines.

The construction is like a linear ring body, which has been placed inside of the base and not

cover the environment off as well aas bring much beauty to this space and landscape. We hope that the building space itself is like a 3-cycle line, which will bring people to experience the fairyland. The spatial layout of this spa is three annular regions basically. They are mainly equipped in accordance with the surrounding landscape. In addition, the floors on the earth are upgraded in line with the surrounding terrain height.

The north hall, +0.2m /restaurant and swimming pool, +0.5m /South foyer, +0.8m / suits, +1.1m

Three annular regions:

No.1: hall, cafe, spa and the internal swimming -water area

The space is surrounded by water, giving you a feeling that float on the water. We hope the closing and extending relationship can combined the space and surrounding waterscape together. At the same time, due to the spring will generate mist in winter, it will provide people a feeling in fog phenomenon.

No.2: restaurant-flower

The restaurant is close to flower area. The external part of the room is flower slope. The dinning place and the room balcony are taking cherry as the theme. We hope to use the flower and landscape as the theme to shape the delicate atmosphere of restaurant.

No.3: Suits-foerst area

The room area is in the forest, which space is surrounded byu bamboo. Choosing bamboo to create the forest is because that the bamboo fores is in line with this space properties. Most important is that it can form a natural high-density protection belt and anti-cold feature can provide rooms a basic privacy in four seasons. Particularly for the spa areas of the internal room.

We hope this resort is a place that let people feel the environment in terms of vision, hearing, smell and touchable.

It is like to hear the sound of water, birds singing and sound from bamboo , to feel the hot spring and fog, and smell the flowers.

/凰茶会

一提到茶楼,留在人们脑海中根深蒂固的印象就是灰砖、青石、木雕、木格栅等浓重的传统中式符号,本案的出发点就是颠覆这一传统观念,呈现给客人一个别样的空间感受。

凰茶会是一家专业经营高等普洱茶的公司,为了配合茶品颜色、口感上的特点,设计上将空间的主材和主色调定位在非常素雅的材质,如亚麻布、木纹石材,这些非传统茶楼所用材质营造出了一个素雅的环境大背景。但是作为茶楼,千变万化不应脱离其中式的根本风格,那么我们就通过点的形式来体现其中式风格的筋骨。

既然是要颠覆其传统概念,那么我们就在其材质、工艺、表现形式等多方面来进行重塑。

首先,大面积使用的亚麻布这一非常规材料,并对其施工采用侧角硬包的形式,突出其粗料细做的工艺性;

其次,本案大量使用了紫铜这一材料,其材质本身就给人一种奢华而又内敛的特性,在地面应用上将其与石材拼接,使特性发挥至极致,使客人行走其上有一种别样的体验;

最后,其他的中式元素,如家具、屏风也运用现代的不锈钢工艺,将其不一样的中式风韵体现出来。

整个空间奢华而不张扬、中式而不陈旧、素雅而不平淡,感觉宛若唐诗宋词里的平仄、对仗,水墨里的疏、密、留白。动静、大小、远近、虚实,一起延续着古典美学的传奇。

项目地点:石家庄市槐安路
设计单位:大石代设计咨询有限公司
设计师:张迎军、张京涛
建筑面积:750 m²
主要材料:木纹石、亚麻布、紫铜
竣工时间:2011.07
摄影师:邢振涛

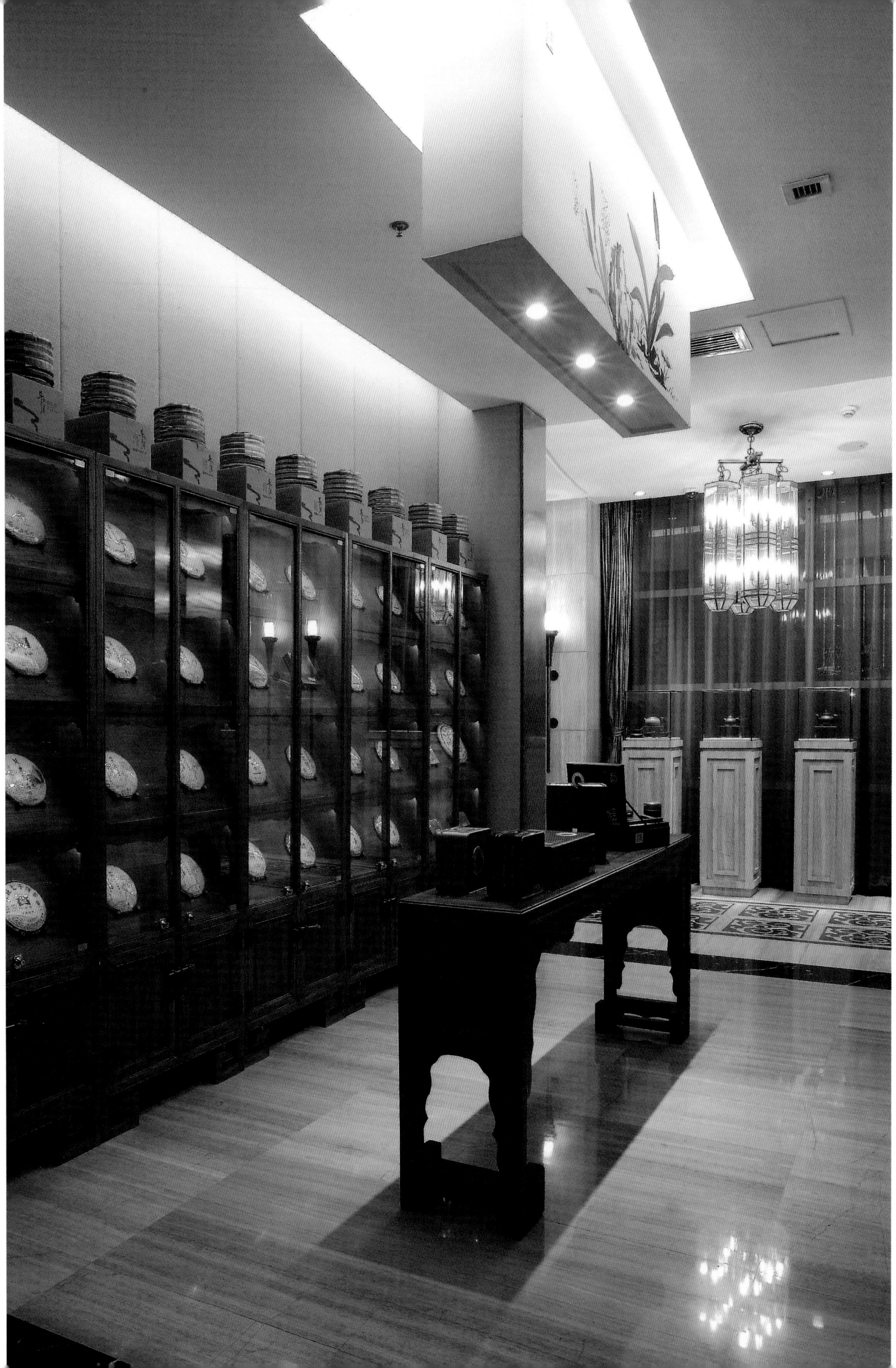

Phoenix Tea House

Once mentioined the teahouse, people would remember that the gray bricks, bluestone, wood carving, wood grille and other tradition Chinese symbols firstly. The starting point of this case is to subvert the traditional concept, showing a different kind space sense to people.

Phoenix Tea House is a professional tea company on high-class Pu'er tea. In order to cope with the color of tea, as well we its taste, the space materials and the main color are all positioned in the very elegant materials in design, such as linen, wood-texture-like stone. These materials selected by this non-traditional teahouse have created a simple and graceful background. However, as a teahouse, the changable appearance can not depart from its fundamental style. Then, we adopt the general points to reflect its basic style.

Since we would subvert the traditional concept, then we will re-shape the space on its materials, process, and expression forms.

FIrstly, it adopts linen by a large area and takes hard package forms to show its process of coarse material.

Secondly, this case has adopted much purple copper. This material itself gives us a luxurious but humble characteristics. For the ground, it connects with the stone to play its features fully, allowing guests to walk on a totally different way.

Finally, the other Chinese elements, such as furniture, screen all adopt modern stainless steel technology, showing its charming Chinese mood.

The whole space is luxury but humble, Chinese style but modern, graceful but not boring. You will feel you are reading a poem of Tang and Song Dynasty, The sparse, dense, blank of the ink picture, quiet or active, big or small, far or near, abstract or reality are all the extention of classical aesthetics.

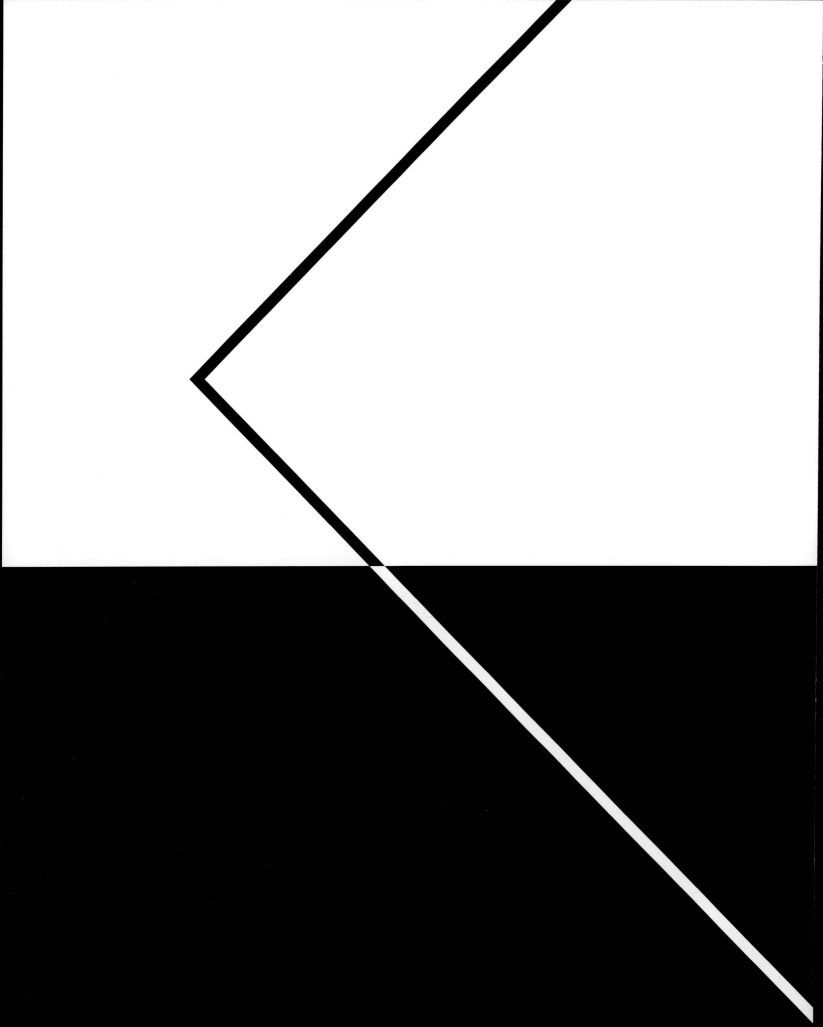

REGION CENTRAL CHINA
华中

郑煤仁记体检院

项目地点：河南郑州
设计单位：郑州弘文建筑装饰设计有限公司
设计师：苏四强、王政强
参与设计：张雷、段素华
建筑面积：**5000 m²**
主要用材：人造石、壁布、彩色玻璃、白木纹石材、涂鸦
摄影师：周立山
撰文：苏四强

柔和安静的曲线造型徜徉在空间中，消除了前来体检者对医疗检查的恐惧感。明快的色彩使体检者能快速、清晰地找出所处位置及各个体检区域：负一层体检区的男区天花选用清新、宁静的蓝色，女区天花选用亲切、优雅的粉色；二层VIP区天花选用了欢愉、温暖的橘黄色；而三层门诊区则选用了明亮干净、雅致的白色。

使用跨界的设计手法（把平面设计运用到室内设计中）来丰富空间，趣味性的手绘涂鸦艺术化了空间，蓝天白云的梦想、环保的理念，游弋在天花造型中好似一扇扇开启的天窗，不同色彩的区域会看到不同的"画面"，巧妙地隐喻了与自然的相连关系，将清新的空气"引入"室内，使原本不高的空间不再压抑而是更加通透。

来到这里是一种美妙的健康体检之旅，正如我们客户的经营理念一样，仁人体检带来五彩缤纷的美好生活。

Renjihealth Checkup Hospital

Soft and quiet carved shape fulfill this space, eliminating the sense of fear of the exminator. The bright color can let the examinator to locate his position quickly and clearly. At B1, it is male area, which ceiling is selected clearly, quietly blue color. The female ceiling is selected friendly and elegant pink color. At the second floor, the VIP area is adopted joyful and warmly orange. The third floor is selected bright clean and elegant white color.

The intercross design way (put the plain design into the indoor design) to decorate the space. The funny hand painting has made the space nicer. The blue sky and white cloud represent dream, environment-friendly concept, patterns stamping on the ceiling seems like various skylight. Different color area will find different "Pattern". They have connected with nature cleverly, letting the fresh area into the indoor which will make the low space seems more transparent.

It is a nice health examination trip to come here, just as our operation concept, Renjihealth Checkup Hospital will bring you a better life.

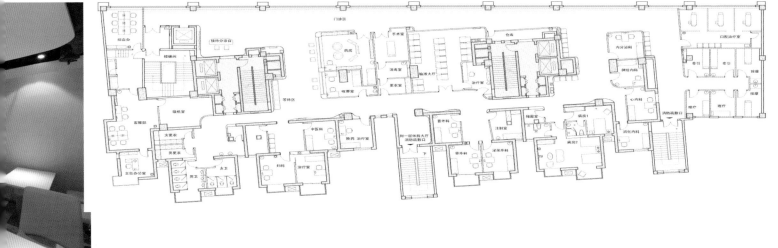

武汉别墅样板房（B1户型）

项目地点：武汉
设计单位：壹正企划有限公司
设计师：罗灵杰、龙慧祺
摄影师：罗灵杰

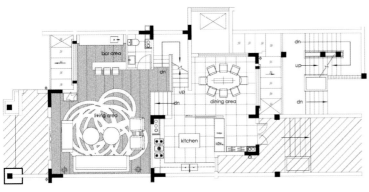
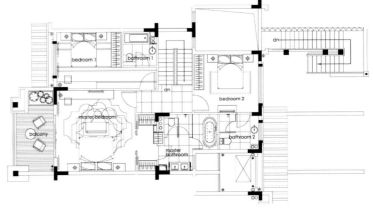

　　这所别墅样板房以"立体花线"为设计主轴，对称的花线设计是欧陆式设计不可或缺的元素，但在天花、墙身，以及地脚位置加上对称花线的设计又过于传统，因此设计师为"花线"注入全新的元素，使之变得立体化，更让立体花线散布于整幢别墅的不同位置中，这种创新大胆的设计理念，提升了视觉效果之余，更让整幢别墅样板房充满生命力，华丽高尚的格调打破传统浮夸的欧陆装潢，营造出充满时尚感的一所别墅样板房。

　　整幢别墅以黑、白、红为主要颜色，客厅及饭厅是以白色为主，除了用上白色云石外，于天花及墙身设计了不同长度的立体花线，设计师未就花线的位置定位作刻意对称的安排反而更显空间感；加上巧妙地利用花线的线条组合而成灯壳，设计既简约又时尚。楼梯把手的设计配上造型特别的椅子家具，不但贯彻别墅样板房西式的设计风格，更带出高尚的王者气派。另外饭厅中的屏幕，因为用上白色透视屏幕，重叠花线砌成的图案，打破一般墙体的压迫感。

　　主人睡房的特色墙是另一设计亮点，此特色墙身是利用大量的切割木板拼凑而成，堆放一片片的立体花线木板而成为一面完整的花线壁板，当中更设有固定壁灯及床头桌的设计，非常特别。设计师考虑到整幢别墅样板房布满立体花线后已足够丰富，其他的家具都以简约线条为主，避免花巧或造型过于复杂的装潢影响整体设计效果。

　　至于厨房的设计，也因为强烈线条感的花线设计已经非常突出，因此设计师选用了黑色，配合黑色的云石墙身，同时厨柜、台面亦以黑色为主色，凝造出非一般的型格。

Molding Show House

This show house is inspired by "molding". Molding is one of the important elements in the continental style, but it becomes too tradition to use molding in the ceilings and walls nowadays. As the result, designers input new elements and spread the molding all over the whole show house. With this innovative idea, not only enrich the visual effects and its vitality, but also enhancing the modern and grand style as well.

As the molding lines effect is strong, only white, black and red being selected as the theme colour. Living and dining areas are in white color, in addition to the white marble, there are molding in various length and sizes on the ceiling and walls. The molding being located seems-randomly-but-well placed and enhancing the spaciousness. Furthermore, to interpret the simple but stylish design, designers make use of the molding to become the lamp shade. Stairs handles and the unique furniture are well interpreting the western interior design style.

In addition, there are custom made screens in the dining area. To replace the curtain, designers make use of this white and see through screens, with the repeating molding pattern, to enrich the comfortable and relaxing atmosphere.

Another highlighted sight point is the feature wall of the master bedroom. It is made up of ample laser cut wood panels, piled up to form a three dimensional molding panel with built-in wall lamp and bedside table. To ensure this extraordinary molding panel being the chief attraction and avoid fancy décor affecting the overall design, the furniture is in simple shape and design details.

Similar to the living and dining areas, with the strong molding lines features on the cabinets and countertop, designers simply use the black marble to match with the theme color for the kitchen. Without using too much color scheme and complicated design details, this black color kitchen is really in cool style.

106

华中 REGION CENTRAL CHINA | 115

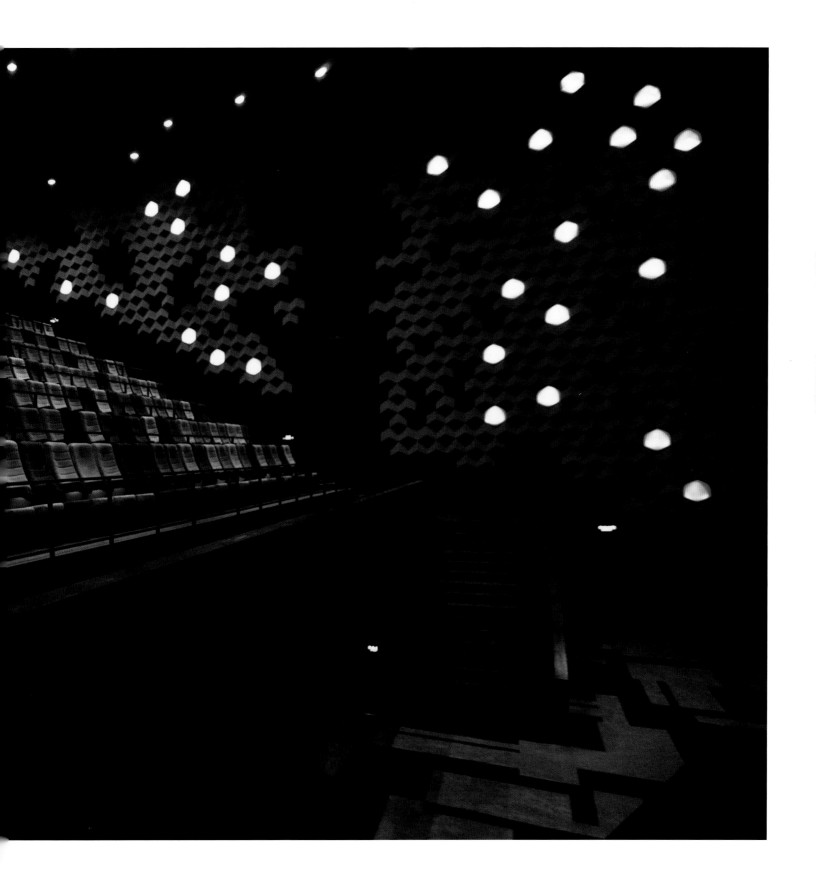

武汉畅响会所

项目地点：武汉市武昌区解放路518号
设计单位：阔合国际有限公司（www.cvox.com.tw）
设计师：林琮然
参与设计：李本涛、林盈秀
建筑面积：2000 m²
主要材料：木工、白色石材、玻璃、不绣钢、马来漆
摄影师：黎威宏

武汉水多,两江交汇,百湖纳于其间,养成了武汉那自由包容的内在个性,武汉男人的豪爽与女子的泼辣,形成了这城市天生有的江湖底气,如此的大雅与大俗就这样在武汉交织。设计一个武汉新时代精神为主体的Echo Club畅响会所,必须包含着武汉特属的"自由休闲"精神。设计之初建筑设计师林琮然,走访了旧武汉的街头,那生猛的大排挡与露天麻将,融合了叫卖声、麻将声与浓厚的笑声,自然在他脑内形成了一种忘我音感,空间的想象就来自于"余声绕梁,音浪荡漾!"他希望打造出音乐进入空间内如同行云流水般的自然完美,利用特殊的前卫表现创造出一种静止的生动画面,强烈的视觉震憾,让人游走其间如同唱到忘我神离,而充满艺术感的基调对比炫丽幻化的灯光,更展现出清新纯真底下那热情澎湃的血液。年轻是二元一体的冲突性表现,流行是新旧交替的过渡,以城市个性为底蕴放入戏剧性与幻想,将伴随年轻基因重现自在的码头。

完美的想象来自于林琮然梦里的白与武汉的红,所以入口蔓延武汉城市的记忆,利用红色打造出一片片曲版,表达出韵律化形变的空间,而镜面与液晶画面的搭配,扩大了空间感,也创造出展厅般的影音,若说起点是红色的火光,那小女孩在火柴棒燃起后看到梦想,就在那偌大白色梦境中展现,火红的动感一瞬间被自由的白给舞动,视觉的想象在大厅内被激荡,音符画过天际转换成五彩幻化的光景,

1. 服务台
2. 上网区
3. 甜品吧台
4. 休息区
5. 礼品展示
6. 录音室
7. 包厢
8. 酒水超市
9. 超市结帐区
10. 备品室
11. 餐厅

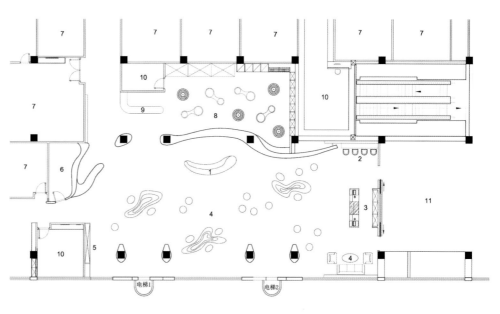

PLAN SC: 1/100

音浪燃烧让地面上象征黑白琴键的石材瞬间弹奏起来,在超现实的空白空间内,自在随意摆放的白色卵石沙发,起来稳定空间的力量,让人在此等待也轻松愉悦,梦里的幻境打破了现实,飘进了平凡,年轻就在多重层次的白色中展现出音乐,而想象力的飞驰是建筑师献给年轻人最完美的音符。

以年轻人为主的消费空间,如何让来到这边的人有高度感受,这是考验着设计师的思想与设计,在高度微博的讯息时代,关注、互粉、@等行为,若可以安排到现实空间内,满足人类天生的好奇心,这空间转化过程就是种未来。所以建筑师不满足大厅只是单纯性的等候区域,巧妙的加入了网吧与个性化录音室,刺激了人们在等待中产生看与被看的关注感,如在录音间的外墙设计了一个玻璃展示窗,窗外并特别留设了一个半开放式的关注等候区,让好奇的人可以透过玻璃窗中的麦克风,在此等候并关注录音间的素人歌手,进而产生互粉与@的多方交流,由实体外部的空间去诱导人们产生虚拟世界的讯息改变,这样的观念不仅丰富了空间的多重样貌,也改变人们共享欢乐的体验模式。

Echo Club 畅响会所充分考虑时尚品味,让武汉的特色与年轻活力在这此地被升华,用艺术与情趣将传统的歌唱空间幻化了灵动的气氛,林琮然努力实践坚持品味、原创梦想感受美好,全力开创新的欢乐境界。

Wuhan Echo Club

Wuhan has much water. It is the crossing part of two rivers and many lakes are focusing here, formed Wuhan's freedom and inclusive personality. The bold and forthrigh of man as well as the generous of woman have formed the basic tone of this city. Such elegant and common are collected in Wuhan. As a new times landmark, Echo Club shall contain the special "freedom and leisure" spirit. At the beginning of the design, the designer Lin Zongran has visites all ancient street in Wuhan. The open restaurant and open mahjong, merging with the selling voice, mahjong voice and laughters, they have formed a kind of sound sense. The imagination of this space is coming from "Sound voice, flowing among the space." The designer hopes to create a natural perfect music space as the flowing clouds and water. Using the special advanced performance, it has generated a static sound of animation face with strong visual shock, letting people immerse into it and forget themselves. While the dreamy light that is full of artistic sense shows a enthusiastic blood under the fresher and purity face. Young is one conflicting performance. Popular is the transition part of old and new. Taking urban personality as base and putting into the dramatic and fantasy, they will reappear on the free dock.

The perfect imagination comes from the white in Lin Zongran dream and red of Wuhan. Therefore, at the entrance, there is full of memories of Wuhan city. Using red to create pieces of part, it has shown a changeable space. Furthermore, the mirror matching with LED screen, they have enlarged the space and also have created a exhibition-like musical. If we take the red flame as the starting point, then the little girl see her dream after igniting the matchstick. Her dream is realized on the huge white dream place. The dynamic moment of red is inspired by white, the visual imagination is stirring in the hall. The musical notes drawing through sky and then converting into colorful landscape, the sound wave is burning on the ground, which symbols of black and white keys on the ground. Among the dreamy blank space, the

white pebbles sofa are placed naturally, showing a strength power. It will give people a comfortable feel here. The dream world has been realized here and come into our life. Young has been performed here for many times. Furthermore, the endless imagination is the most beautiful rhythm the designer given younger.

This consumer space for younger that how to let people here to have a strong sense is the most difficult for designer. In the ear of highly micro-blogging messages, concern, mutual addition, @…and other behaviors, if they can be arranged into the space to satisfy people's curiosity. This converting process is to plant future. Therefore, the architecture does not satisfy the pure waiting function of the hall. He has added web bar and personalized studio in it to stimulate people's mind and heart. For example, there is set a glass display window at the outside door of the studio. There is also set a semi-open concern in the waiting area, letting curious people can see the microphne through the glass window and concern the singer in the studio. Then, they may concern with each other and communicate with each other. Through the entities external space to the virtual world for information, such concept has enriched space appearance and changed people's entertainment.

Echo Club has fully considered the fashion sense, letting the Wuhan's feature and young people her can be improved through art and interesting. They have enriched the traditional space with live elements. Lin Zongran is alway trying to adhere to original dream to open up a new joy realm.

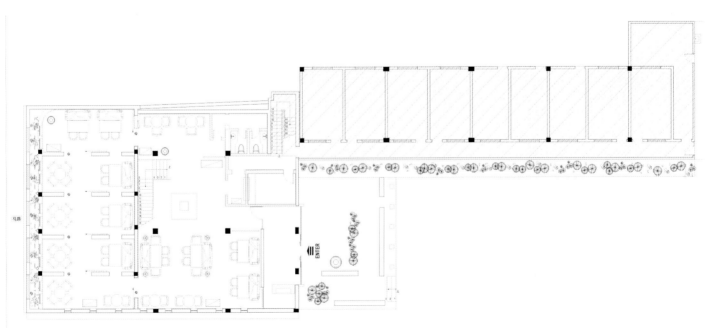

域——中国室内设计年鉴

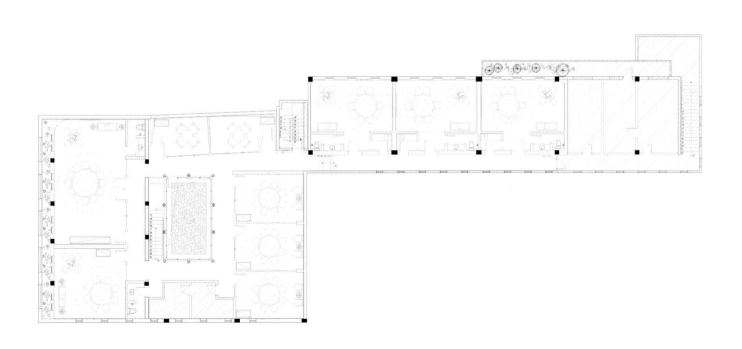

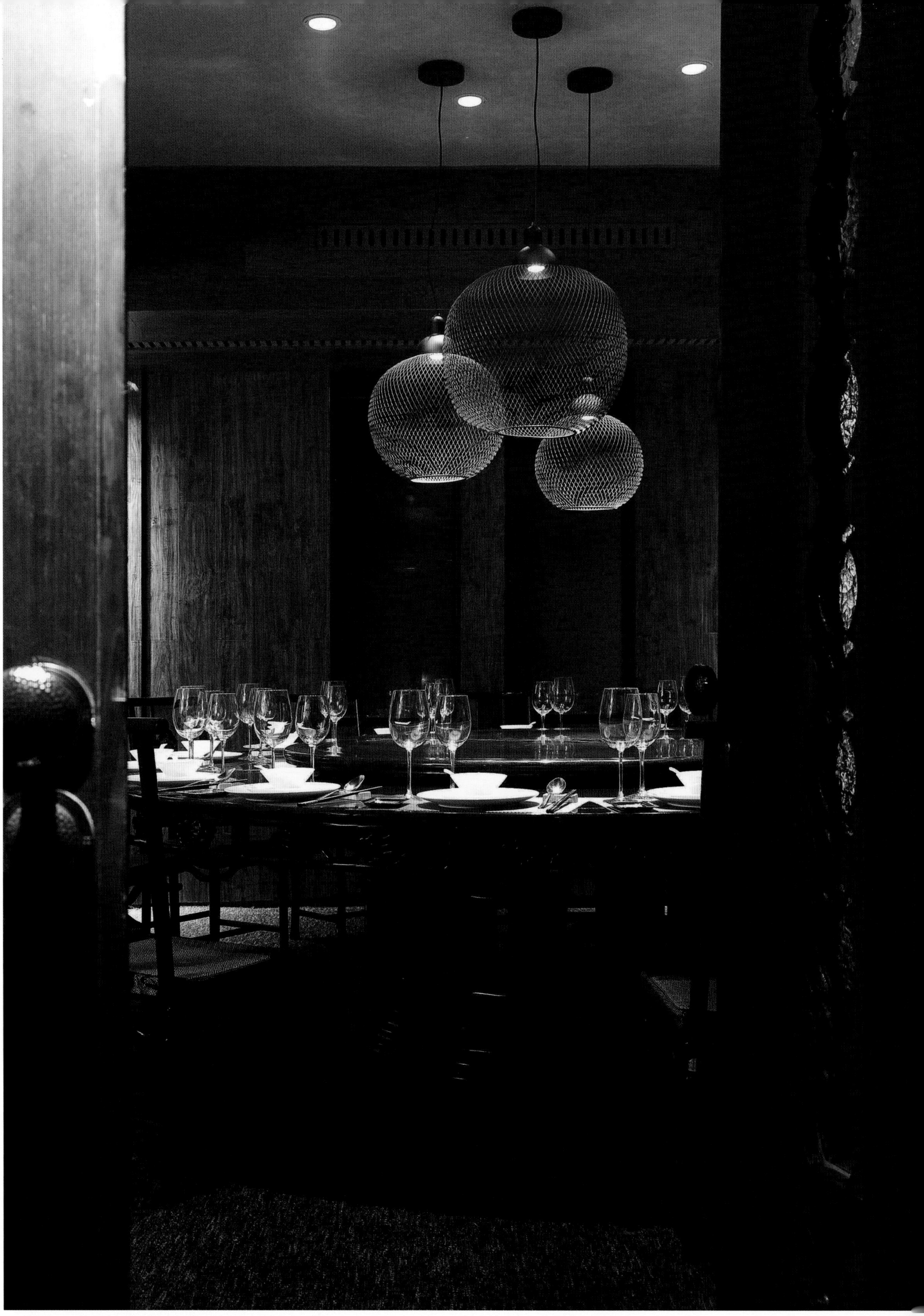

光耀候鸟高尔夫球场拥有极好的自然资源,即南中国最大的候鸟栖息地。

候鸟球场里采用小型别墅风格,依山环湖而建,平层别墅设计,自然中带着奢华的味道。这个别墅以现代中式风格为主旨,简约澄净,比例与线条,色彩和纹理,皆与自然融合。充满宁静含蓄的气氛,贴近人内心深处,体现清净与简单平等的风格。简约的处理手法加入细节上的处理,也避免过多的陈设而导致无谓的混乱。色彩方面使用相同柔和色调子达至色彩的和谐,加上深色调的木线纹理营造对比有趣的效果。宽阔的水池庭院和自然景观,令整个环境写意又豪华。

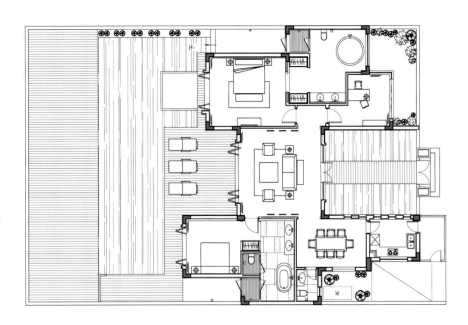

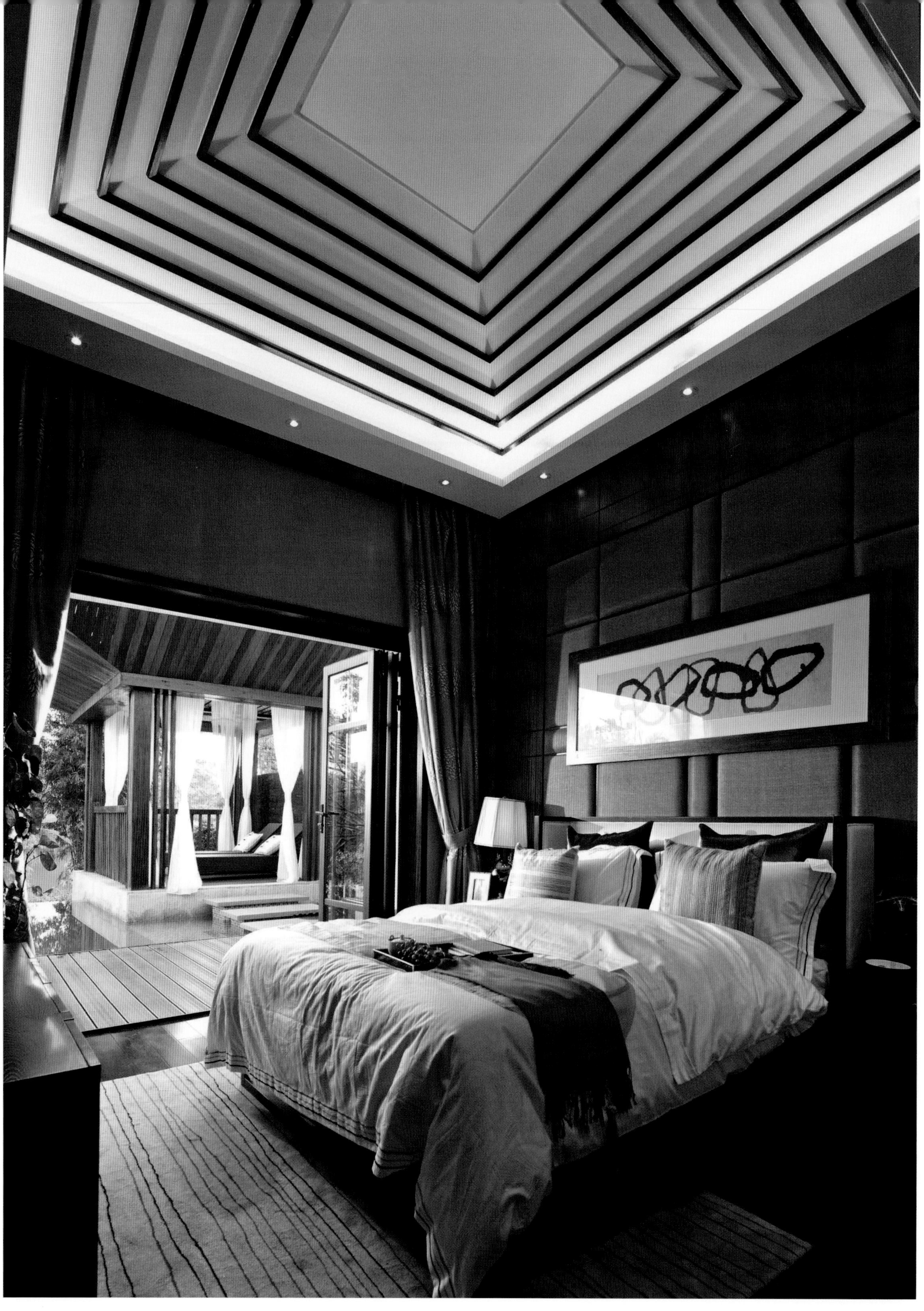

Chinese Style Villa in Guang Group Bird Gold Course

Guangyaohouniao Golf Course has excellent natural resources. It is the largest migratory bird habitat of China. The small villa in this golf course, building around the lake and depending on the mountain, adopting the plain villa design, shows the luxury taste in nature. This villa takes modern Chinese type as the major tone, simple and purity. Its proportion and lines, color and texture are all mingled with nature. The quiet atmosphere has gone into people's heart. The simple treatment skill combined with the details management has avoided the fussy layout. The designer adopts similar soft color to achieve the harmony in terms of color. Compared with the dark wooden texture, they have shown a funny effect. The widen water pond yard and the natural scenery make the whole environment elegant and gorgeous.

深圳 La Vie Pub

项目地点：深圳福田购物公园
设计单位：厦门东方设计装修工程有限公司
设计师：吴伟宏
建筑面积：**700 m²**
主要材料："泰斯特"户外硅藻泥、水泥金刚砂地板

　　本案位于深圳福田CBD中心区，原先是一个荒废的屋顶，通过设计整合，变废为宝，创造了一个时尚浪漫的休闲空间。设计师在空间布局上突破传统布局，使人与人之间的互动更加密切和方便，这是这个PUB亮点之一。在都市中工作，人的压力大，所以本案在设计选材上均采用环保材料，水泥地板、硅藻泥、原木、干枝。全场没有空调系统，而是采用水雾化系统降温，灯光LED控制系统，给都市人带来心灵的归宿。

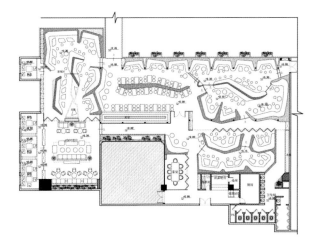

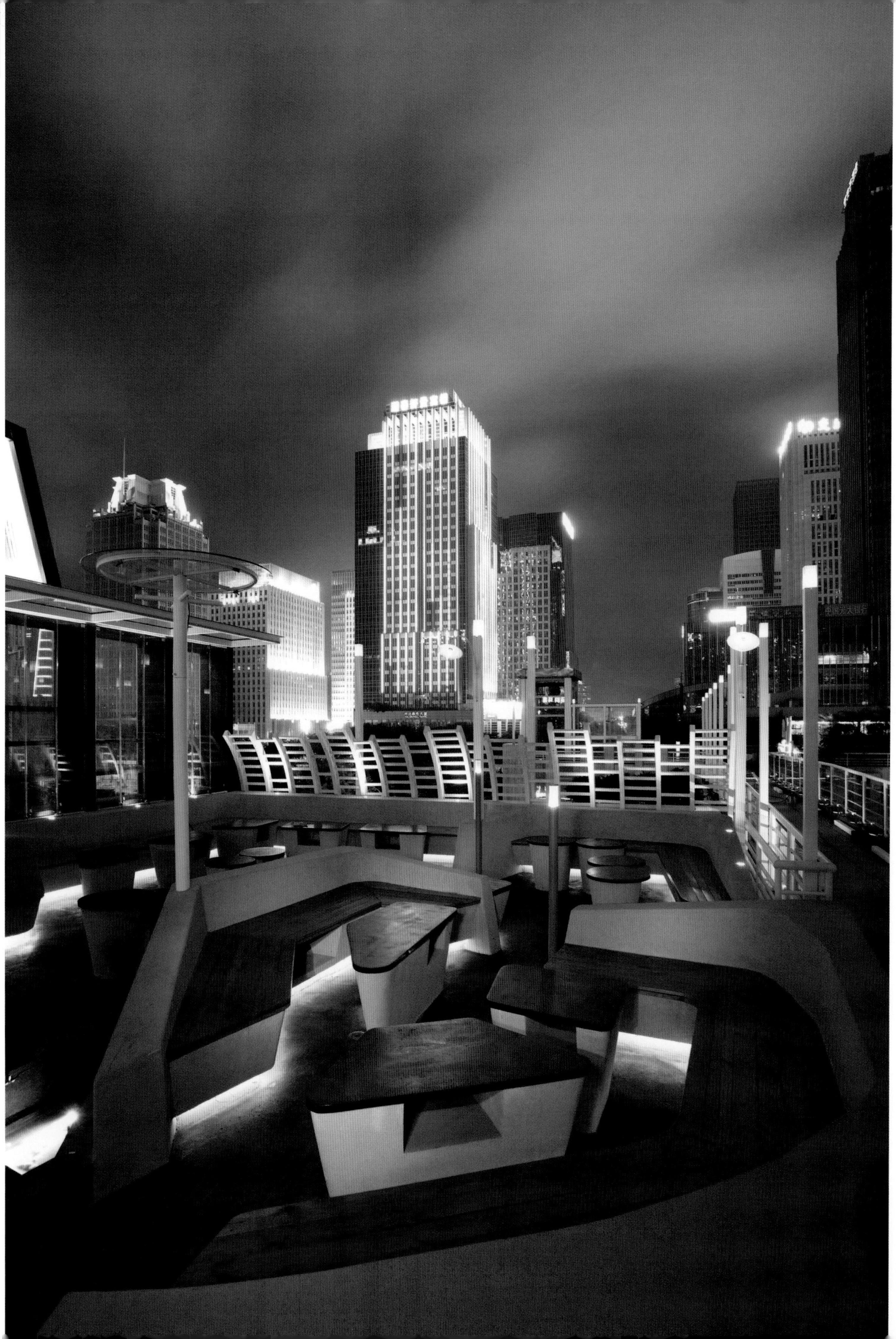

Shenzhen La Vie Pub

This case is located at CBD center part of Futian, Shenzhen city. It was an abandoned roof. Through design, it has converted to treasure from waste, creating a fashing and romantic leisure space. The designer breaks the traditional layout and let the interaction between people more closer and more convenient. This is one of the highlights of PUB. Urban people all have big pressure, therefore, this case adopt environment-friendly materials, cement floor, diatom mud, logs, dry branches. There is no air-conditioning system. It is adopted water atomization system for cooling. The LED control system will bring a spiritual home to the urban people.

西帷办公室

项目地点：深圳
设计单位：汉诺森设计机构
设计师：王文亮
建筑面积：**750 m²**
主要材料：瓷砖反面涂刷、白色氟碳漆金属网
项目时间：**2011.02—2011.07**

西帷是一个全新成立的品牌。作为中国企业由生产化向品牌化转变的一员，西帷希望建立起具有国际化视野的办公空间，来接待国内外的代理商和客户，传达其进步的企业形象。空间的整体设计上，我们注重传达空间带来的品牌感受。通过"浪费"工作面积而换取的一条公共长廊，作为步入企业内部空间的过渡，提升人们的情感预期。进入办公空间，我们用白色铁网的半封闭隔断，在紧凑合理的规划中带来明净舒畅的工作氛围。材质上我们反对堆砌，首次尝试利用廉价瓷砖的反面肌理通过重新拼接打磨、涂刷，显现出一种精致、有序的美感，温馨而时尚。墙面与白色金属网的连接呼应，为空间注入纯净而轻盈的独特气质。

INCANA OFFICE

INCANA is a new brand. As member of Chinese enterprises converting from production to branding, INCANA hopes to establish a international level office to recept the agencies and customers from both home and aboard to convey its image. For the overall space design, we focus on conveying the brand sense brought by space. Through "covering" working area, a long public corridor is gained. It is a transmission part for entering the internal space, increasing people's emotion expectation. Entering this office, we adopt white iron net to form a semi-closed separation, which has brought a clean and comfortable working atmousphere by the cohere and reasonable design. We don't recommend pile materials together. We try to selecte cheap tile's back face to show a delicate, orderly, warm and fashion after re-polishing, painting firstly. The wall is reponsed to the white metal net, which has added the unique character of purity and lightful to this space.

易菲展馆

项目地点：深圳
设计单位：汉诺森设计机构
设计师：王文亮
建筑面积：**600 m²**
主要材料：一次性水杯、纸
项目时间：2011.05-2011.07

色彩上选用自然米白，纯净而柔美，馆体宏大绵延，材质轻盈温和，通过人们习以为常的传统物料，巧妙重组形成非常独特的建筑"皮肤"，给人过目不忘的视觉特征，借以重新唤起人们对"通常"这一习惯的思考与颠覆，从而达到与企业品牌文化所传递的精神理念的高度契合。

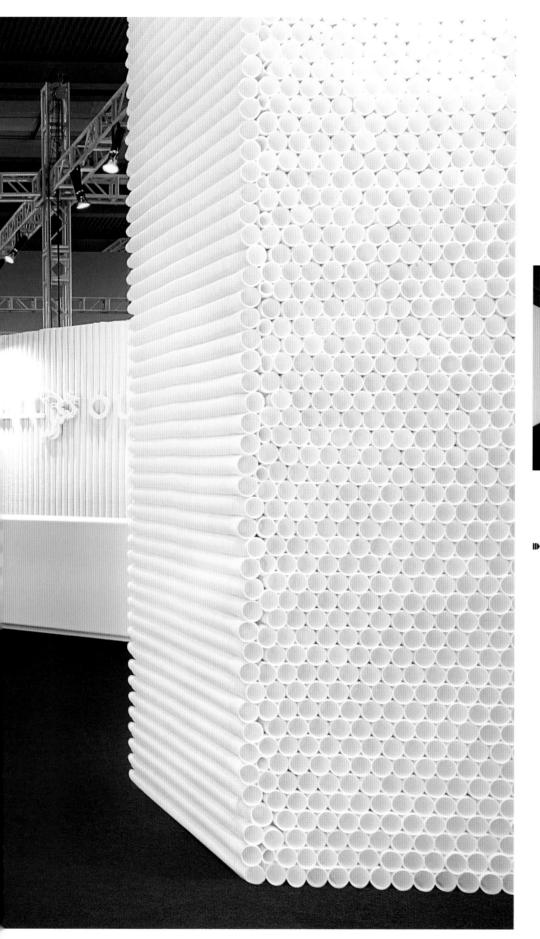

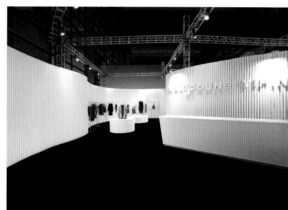
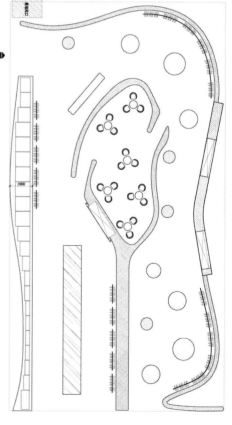

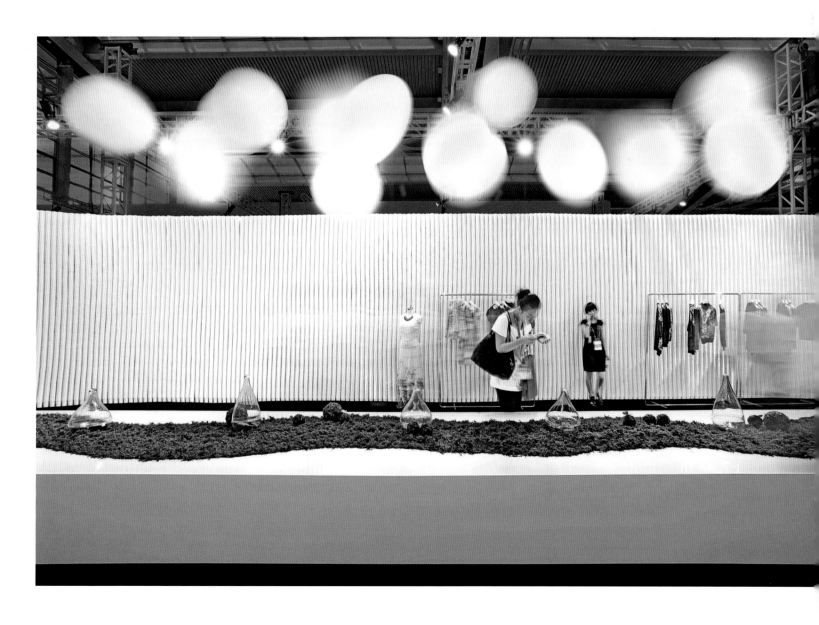

YIFINI Exhitbition

It is selected the natural white color, softly and purity. The main body of the pavillion is grand and long. Its material is light and gentle. Through the traditional materials, the designer combined them together into a very unique construction "Skin", giving people an impression. Thus, it re-call people's consideration and love of the so-called "traditional" so as to achieve the highly identical in line with the enterprise brand culture spirit.

雅士阁美伦酒店

项目地点：深圳
设计单位：HSD水平线空间设计
设计师：琚宾
参与设计：张轩崇、许金花、石燕、尹芮、谭琼妹
建筑面积：25000 m²
主要材料：灰木纹 科技木 艺术地毯
项目时间：2011.01.01
摄影师：孙祥宇
撰文：裴子衍

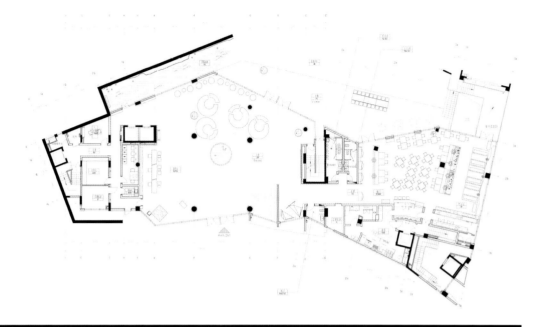

蛇口于南山的关系，就好比南山于深圳的关系。理应是隶属，人们却更多地愿意单独对待。热闹还是一样的热闹，繁华还是一样的繁华，但不杂乱，熙攘中带着点生活味的可爱气息，当然了，还沾着点异域的情调——总之，很特别。

说深圳是海滨城市，要闻海边的气味，还得去蛇口。那里和大梅沙的海边是完全的两个味道，有点潮，风中海味很浓，生活的气息很重，却感觉很"旅途"，很是异样。也许人们一直向往的就是在旅途中行走，在旅途中停留，一路有不断变幻的风景，一路经历着不断的别样精彩。蛇口受欢迎，应该和这不无关系。

美伦酒店的名字同样也很"蛇口"，不论是不是集团、全国有着多少个另外的所在，美伦，美轮美奂，很是适合所处的位置。

近着香港，挨着海边，不远处还有着大南山。看着地理坐标，即使是几间简单的茅草屋坐落于此也是很有意境的事，何况，是这间由招商地产打造的符合新加坡雅诗阁管理公司要求、兼商务与度假为一体的酒店。

风水，有风有水，或者说驻风纳水，中国人

一向都很讲究。刚巧，美伦酒店都符合。建筑是由都市实践的孟岩所设计，远远呼应着周边的建筑群落，显得既考究又不突兀。由水平线的琚宾主笔的室内设计，延续了酒店外观的考究建筑感。

酒店大堂天花的造型，保持了建筑外立面形体的一致性，蜿蜒而折回，柔软而硬朗，似无序而协调。木状的铝质材料，在突显整个空间的厚重感和品质感的同时，保持了轻松感和舒适感。体现出了商务与度假的主题。

由灰木纹大理石铺就的地面，兼顾了木与石双重效果，满足了视觉的享受也让人产生一探究竟的好奇心。华丽却感受不到张扬。

同样是灰木纹的大理石墙壁上，是镜面不锈钢材质的艺术装置。与底下的木桩与蜡烛相互映衬。不远处的木头装置，以及再不远处的铜制莲蓬，在搭配酒店整体色调的同时，散发着自然的气息。

大厅墙壁上的木质漆艺装置是仿制的艺术家苏笑柏先生的大作，红漆，斑驳，老旧，浓烈。悬挂着的覆漆瓦片，无序，大气，同样也是中国味的装置艺术的体现。蛇口的阳光洒入，在两件作品之上和之下投下更加斑驳的、或明或暗的影子。东方得特别现代。

无论是环形的沙发还是明式的 Shell Chair，抑或是前台的聚酯椅，传递着东方的信号，同时也呼应着酒店的主题。这是中国文化的别样解读。中国元素丰富而不呆板，处处充盈着灵动的气息。

客房空间的风格与酒店大堂相呼应。很值得一住。比较有特点的是吴冠中水墨画为内容的地毯，让整个空间浸入诗意的氛围，清雅而不造作。

别人说好的酒店要"宾至如归"，那只是说舒适性的。商务、度假酒店的要求则更高，是要既满足舒适性又要求实现体验性的，同时又要兼顾品质。可以不夸张地说，这一切，美伦酒店借着设计师的妙手，都巧妙地实现了。

Yashige Elegant Hotel

Shekou to Nanshan is like Nanshan to Shenzhen City. It shall be subordinated relationship. However, people like to treat it separately more. Excitment as usual, luxury as usual, but they are all in orde. Among the hustle and bustle, life is still sweet and cure. Of course, there is exotic mood also--in short, very special.

If you mean Shenzhen is a coastal city and you want to enjoy the coastal sense, then you'd better to Shekou. It is totally different from Dameisha beach. Shekou is a little bit moist, with heavy sea flavor in the wind. Even there are much life sense, you will still feel that you are "On the way", which is very different. Maybe people have been longing to walk on the way, stay on the road to enjoy the ever-changing landscape and experience the various colorful road. That's why Shekou is popular so much.

Elegant Hotel name is also very "Shekou", no matter whether it is group,or how many branches it have in China. Elegant, refers to beauty and nice, which is suitable this place much.

Close to Hong Kong, next to Seaside, there is a large Nanshan in not far distant. Looking at its geographic coordinate, even sveral simple huts locating here is also a nice scenery. Not more say the hotel merging business and holiday inn into a single whole, which is created by Singapore Ascott management company.

Geomancy, refers wind and water, or reserve wind and inhale water, which is cared by Chinese people very much. It is happened to Elegant Hotel. The construction is designed by Mengyan,who has many urban practice. The building is in line with the surrounding buildings, which seems elegant and lofty. Designed by Jubin, expert of horizontal line, this hotel has succeeded the architectural sense on appearance.

The shape of bolly ceiling, has kept the consistency of the building appearance, winding and turing, soft but powerful, which seems disorderly but coherence. The wooden-like aluminum material has highlighted the heaby feeling and a sense of quality of the entire space as well as maintained a sense of relaxation and comfortable. It just reflect the theme of business and holiday.

The floor made from gray wood-texture-like marble has taken into account the dual effect of wood and stone, satisfying people's visual requirement as well as curiosity. It is gorgeous but humble.

Same from gray wood-texture-like marble wall, there is a mirror stainless steel art installation, refecting with the wooden pegs and candles. The wooden device, as well as lotus seedpod in near distant are spreading a natural atmosphere coupled with the overall tone of the hotel.

The wooden lacquer in the wall of the hall is an imitation of artist Mr. Su Xiaobo's great work, red paint, mottled, oldish, and passion. The suspended painting tiles are disorderly and grandness, which is also full of Chinese style. When sunshine pouring into Shekou, these two masterpieces seem

more shadowy from up to down. It is the represtation of modern Oriental.

No matter the circular sofa or the open Shell Chair, or the polyester chair in the reception, they are all conveying the signals of East as well as responding with the hotel's theme. It is the different interpretation of Chinese culture. Chinese elements are richful but not rigid, full of smart mood everywhere.

The room space style is in line with the hotel lobby, It is worth living. The most different one it the carpet that takes Wu Guanzhong's drawing as the pattern, letting the whole space immerse into the poetic atmosphere, elegant but not pretentious.

It is said that a good hotel shall be "Home equivalent". That's only refering to the comfortable. For the business and holiday hotel, it shall satisfy the comfortable requirements and the experiential as well as the quality. It is no exaggeration to say that through the designer's hands, Elegant Hotel has realized the above requirements all.

与自然共生

项目地点：佛山桂城怡翠馨园
设计单位：佛山尺道设计顾问有限公司
设计师：李嘉辉、杨铭斌、何晓平
建筑面积：**230 m²**
主要材料：木饰面、乳胶漆、玻璃、木纹砖、不锈钢
摄影师：李嘉辉、杨铭斌

在原有直板空间中利用建筑的"反形"思维,将空间布局重新划分,加入不规则的线性涂料作为主要的墙面材料,除了令空间更富有层次和动感外,还具有与自然共生的感觉。每一个弧度带向另一个不同的景观,令人在不同的空间感受不同的景观情景,一个景就是一种与自然共生的生活方式。

LIVING IN NATURE

Converting original square space in contrary thinking, we redesigned space with irregular liner painting as the featured wall material. New space looks rich in layering and motion, as well as the life from nature. Every curve leads you to a different view, an from which individual feeling comes and presents a living in nature style.

华南 REGION SOUTH CHINA | 171

招商金山谷工法展示厅

项目地点：广州
设计单位：HSD水平线空间设计
设计师：琚宾
参与设计：姜晓林、叶向阳、吴圳华
项目时间：2011.02-2011.05
主要材料：亚克力、木地板、LED光纤、LED显示屏
摄影师：井旭峰
撰文：HSD水平线空间设计

门锁地板乳胶漆，窗棂墙纸五金齐，展品琳琅全不乱，含义深远显真机……这不是借着韵的装修内容打油诗句，这就是招商地产金山谷的工法样板房里的内容。如果要将装饰一个精品公寓所需的材料或者其演示图片在一个空间中全部展现出来，要怎么去实现？HSD水平线团队巧妙的回答了这个问题。

这个名叫"荷塘月色"的工法样板房，很是有趣。不合常规，却又实现得仿佛天经地义，让人感觉除此之外别无他法。主体是透明或半透明的亚克力材料，伴着灯光，营造出一个似幽静却同时生机勃勃的场所。荷叶状的不规则展台，有着不同大小形状的同时，还有着不同角度的斜面，方便放置不同展物，也人性化地顾及到了参观的视觉体验。仿生荷叶梗的圆柱状支体也有三种规格，因此全部的展台便高矮错落，显得杂而不乱，透出一种无序之美，显得大气而富有装置性。而展台本身是可以发光的，映在产品或产品图片旁的光晕，以及投在展台下方呈圆弧的光圈，都给整个展厅披上了神秘却又富含科技意味的气息。简约，但不简单。

展台的上方是垂吊着的光纤，细腻而柔软，会随着重力垂将下来，尾端自然地微微卷曲，像简单勾勒的美人的发。吊顶的上方藏有光线发射器。届时光会顺着总长度达5000米数量为一千九百余条的光纤滑落，在展厅的上方组成一片仿佛因时间停止而未来得及洒落的雨幕。光纤本身是没有颜色属性的，给予其什么颜色，便变幻为什么颜色。由此变得科幻，变得未知，变得生动。

展台的下方是灰白色的实木复合地板，表面经过亚光的处理，低调地配合着整个展场的主题。荷塘月色，光影和谐，虽然在秋实先生的名篇里"热闹是它们的"，然而总是带着点淡淡的喜悦，意境也总是美的。

工法一词，出自日本。其诠释为工艺方法和工程方法。在中国，工法是指以工程为对象，工艺为核心，运用系统工程的原理，把先进的技术和科学管理结合起来，经过工程实践形

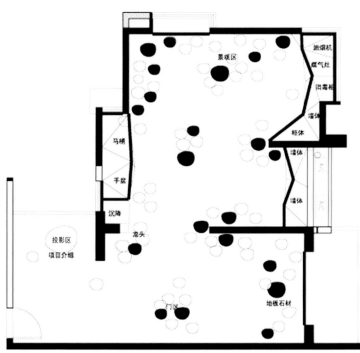

成的综合配套的施工方法。它同时还具有先进、实用、安全、环保、高效率、低成本等特点。这里的"工法",具体到了实物,将工艺方法和工程方法浓缩、提炼,选取最富有代表性和说明性的材料或器物,予以摆放。至于"样板房"的名称,则更多地是在诠释"展示"这一概念。

HSD规划的展览路线应该称得上是一段诗意的漫步,借着灯光的串接,由展品、展台、天花、地板共同组成一个静置的同时又具有漂浮感的行列,从如流水般月光色的纯白到极致的绚丽,从白天到夜晚,在变化的天光与灯光中投射出东方的景致。

这个别致的工法样板在展示房地产项目工法材料的同时,也引领参观者探索居住空间的构造以及表面材料的丰富性,表述着传统意向的同时,传递着自然和人文思想。整个展览以崭新、敏锐的美学观念审视那些工业造物的自然属性,揭示出东方传统的另一面。

Merchants Golden Valley Engineering Exhibition Hall

Locks and flooring emulsion varnish. The window frams, wallpaper and hardware is enough. The exhibits gem is comprehensive. ....This is not the poem about the decoration. This is the sample house content of Merchants Estate Golden Valley EEngineering Exhibition. If you want to show all the decorative materials needed by a excellent apartment or decorate the pictures in a single space, then how to achieve it? HSD horizontal team has addressed this problem cleverly.

The so-called "Lotus Pond and Moon Light" model room is very funny. It is irregularities but looks that right and proper very much. People would feel that there is no other expression way. The major subject is transparent or translucent acrylic materials. Accompanied by light, they have created a quiet but vibrant place. The lotus leaf-like irregular tables have different sizes and types, as wellas different angles.They are convenient for people to display different exhibition object. In addition, it also regards people's visual experience. The biomimetic lotus leaf stem also has three specifications. Therefore, all the tables are located in different ways, many materials but orderly. while the display table is luminous, refecting on the product or product picture,as well as the arc aperture under the display table. They have all put on a mysterious but rich in technology sense, simple, but not boring.

Above the display table, there is a hanging fiber, thin and soft, which can be hanging down by weight. The terminal is curved naturally, looking like the beauty's hair. Above the ceiling, there is hiden a lighr transmitter. At that time, the light will pour dowm along with the over 1900 pieces fiber with 5000 km long. They have created a rain screen on the above, which seems like the time is stopped and the futere isn't come. The fiber itself has no color. It will be colored by what we have given it. Then, it has become fiction, unknown, and vivid.

Below the display table, it is a gray and white wood floor. Its surface has been processed by flat gloss, which is coordinating with the whole space in a low-key way. Lotus Pond and Moon Light, is harmony with light and shadow. Though Mr. Qiushi has written "luxury is theirs", there is some joyous in such beautiful scenery.

The term engineering is from Japan. It is interpreted as a process and engineering method. In China, the engineering means the object of construction. Process is the core. Using the system engineering principles, combined the advanced technology and scientific management, through the construction to form the comprehensive suitable engineering practice. It also has the characteristics of advanced, applicable safety, environmental protection, high efficiency, low cost and so on. The engineering here refers to the physical material. It merges the process and engineering methods to refining and select the most representative and illustrative materials or objects. As for "model room", it is interpretating "exhibition" concept.

The exhibition route planned by HSD can be regarded as a poetic stroll. Resort to the cascade light, the exhibits, booths, ceilings, floors have formed a standing but floating sense. From the snow-white flowing moon light, from the daytime to night, they are reflecting the east beautiful scenery among the nature light and bulb light.

This diligent engineering plate is showing the engineering materials of this estate project. At the same time, it is leading the visitors to explore space structure and the various surface materials, expressing the traditional concept and transmitting the natural and cultural concept. The whole exhibition is taking brand new, keen aesthetic concepts to view these industrial products nature characters, revealing the other side of the oriental tradition.

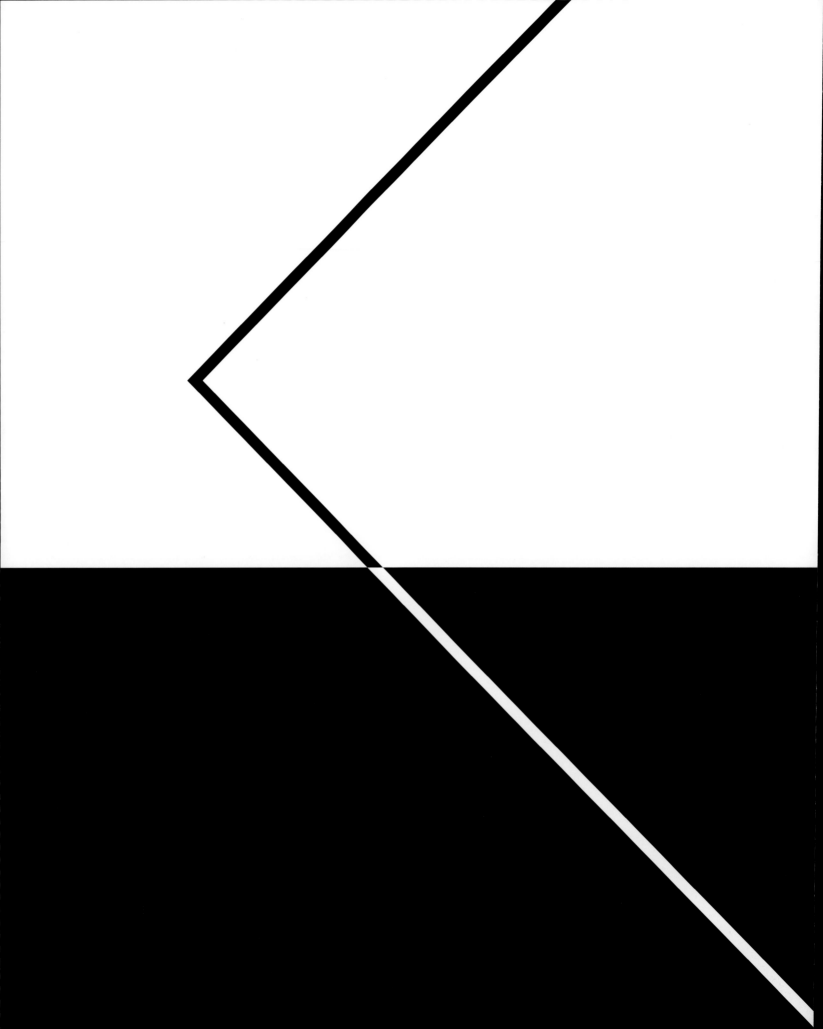

REGION NORTHESAT CHINA

东北

吉林艺术学院艺术咖啡厅

项目地点：长春
设计师：陈旭东、刘寅
主要材料：环氧树脂、不锈钢、钢管、PVC塑料管、尼龙线、爵士白大理石
项目时间：2011

当提笔为此设计写说明的时候,正听到电视报道日本福岛核电站2号和4号反应堆爆炸的新闻。所以就把这个设计标题即兴叫"意象的裂变"吧!这也正符合此设计的初衷。人类文明发展到今天,从农耕社会到工业社会,再到信息时代,我们不禁疑问,文明究竟给我们创造了什么?是得到的多,还是失去的更多?这应该是我们进入下一个时代前每一位设计师所面临思考的问题吧!

吉林艺术学院咖啡厅的设计创意,正是依据自然文明对工业文明的反思制定的创意主旨。"意象的裂变",是精神向物质转换的过程,也是物质细胞体裂变自我完型转化的过程。本案设计力图通过重构的可能,注入物体以自然生命的质量。这种注入证明了物质存在的自然价值,是一种作用于经验下的暗寻快感。而人的经验是有目标的,有动机的,有方向的。它作用于潜意识的懵懂中,每一次体验都是意识与潜意识的重合,从而构成了我们所谓的感觉。

吉林艺术学院咖啡厅的形态控制正是依据这一心理过程的把握,它居于似与不似之间,像与不像之中,有重有轻,有疏有缓,很好地创造了一次崭新的超自然体验。这种体验仿佛把我们置身于未来的意象自然之中,它凝固了我们生命的机体,融化了人与空间的交流,也融化了我们精神所向往的境界。

吉林艺术学院咖啡厅所采用的材料与工艺比较独特。奇异的造型运用了环氧树脂材料,而座椅采用了不锈钢的特制工艺。从某种程度上讲,不锈钢转化了异形椅在通常经验下的物质内涵,它更多的从功能向形式上做了一次大胆突破。而软雕墙则采用了PVC塑料管外缠尼龙线再刷丙烯涂料的方法,使得整个空间在视觉上得到更加愉悦的刺激。室内也大胆保留了原建筑部分墙体与新构筑的艺术形态形成新旧反差。同时也更加突出了雕塑般的流体结构。整个室内空间完全没有老建筑改造的痕迹,最终的艺术效果得到大家的一致认可。

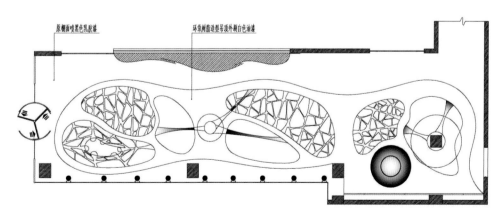

艺术学院咖啡厅天花图

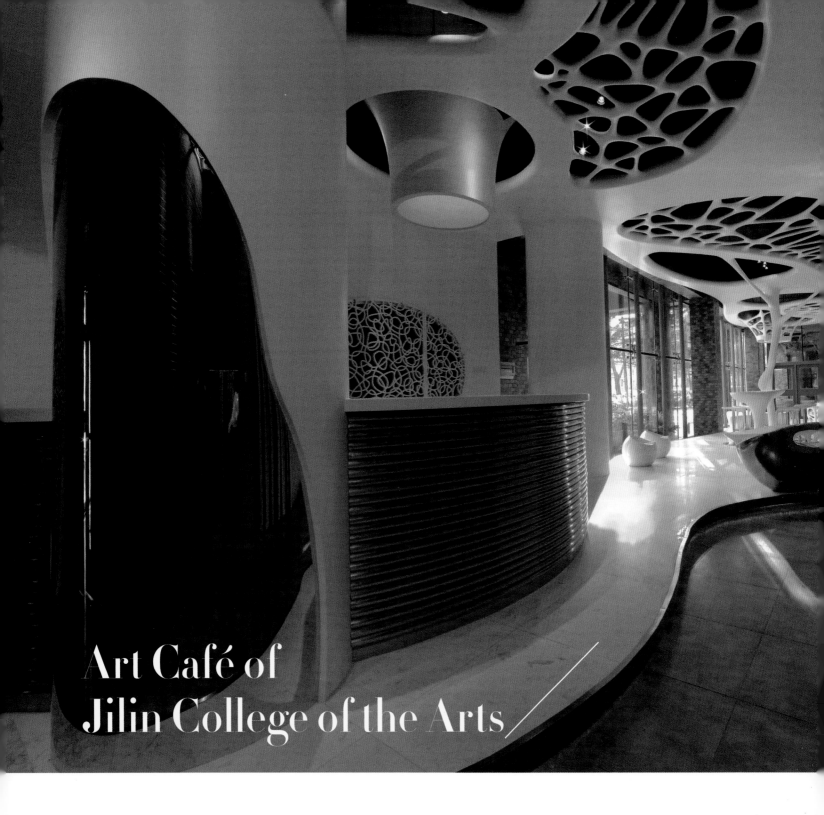

Art Café of Jilin College of the Arts

When picking up the pen to write this instruction for the design, I was listening the TV reports of Japan Fukushima No.2 nuclear reactor and No.4 reactor were bombed. Therefore, I named this design title as "Fission of Image" simultaneously! It is also in accordance with the intention of this design. Development of human civilization today, from the agricultural society to industrial society, as well as to information age, we can not help doubting that, what has the civilization created for us? Whether we get more or lose more? This may be the question that all designers shall facing before entering the next era!

Art Café of Jilin College of the Arts is designed based on the self-examination of the natural civilization to industrial civilization. "Fission of image" refers to the transition from spirit to material, it is also self-completion process through material cell fissile. This design focuses on the possibility of overlap to put natural life in the objects. Such injection has proved that natural values of existence. It is a hidden pleasure under the performance of experience. However, human experience is targeted, motivated and have direction. It acts on the subconscious movement. Each experience is the coincidence of consciousness and subconscious, which constitute the so-called feeling.

The shape control of Art Café of Jilin College of the Arts is based on grasping of soul heart. It lies between similar and dissimilar, like and unlike, bordering light and heavy, loose and slow, which has created a brand new nature experience perfectly. Such experience seems to bring us into the fu-

ture nature scene. It solidifies our body, melting the exchange between people and space, as well as the spirit realm people dreamed.

The material and process adopted by Art Café of Jilin College of the Arts are rather unique. Strange shape has been adopted the epoxy material, while the chairs are adopted special process of stainless steel. To some extent, the stainless has transferred the material meaning of strange chair under normal experience. It is more than a bold breakthrough on the functional aspects. While the soft caved walls are adopted the way of wrapping nylon lines on PVC plastic tube and painted propylene, which had expressed a more visually stimulated pleasure of the whole space. It is boldly that retains part original wall, which forms a contrast between old and new. At the same time, it has highlighted the sculpture-like fluid structure much more. The entire interior space has no signs of renovation on old building. The ultimate artistic effect has recognized by everyone.

亚布力 Club Med 冰雪度假村

项目地点：黑龙江亚布力
设计单位：Naco Architectures 纳索建筑设计事务所（www.naco.net）
设计师：Marcelo Joulia
建筑面积：30000 m²
摄影师：Jonathan Browning

作为国际度假集团 Club Med 在中国的首个度假村，由纳索建筑设计事务所设计的 Club Med 亚布力冰雪度假村提供包括住宿、餐饮、交通、娱乐等"一价全包"的服务，酒店内热火朝天的娱乐活动及室外冰天雪地的景色让客人体验何谓冰火两重天。

在这个位于黑龙江省东北部的地带，雪季几乎占了全年的一半，冬天极度寒冷。才下午四点多，亚布力已经入夜，在黑夜之中，度假村酒店的楼群犹如一座冰雪城堡。

Club Med 提供舞蹈课程、瑜伽训练、有氧操、乒乓球、桌球、麻将、蹦床以及各种为儿童设置的娱乐项目，甚至还有你根本没想过会出现在酒店的空中飞人马戏课程。除此之外，酒店还为喜欢夜生活的人提供了一个热闹的场所——森林酒吧。酒吧的整体设计以亚布力所在的五花山为灵感，家具采用山上的五种色调，其间错落分布着树皮壁纸覆盖的立柱，专门定制的地毯图案来自五花山地表图，而吧台则是对传统中药橱柜的重新演绎。夜晚的森林吧是一幅热火朝天的场面，客人不但可以免费品尝各式调酒，还能唱卡拉 OK，或者在服务员的带动下一起跳舞。当然，对很多客人来讲，来到亚布力就一定要和寒冷交战，那么露天按摩浴池就是个直面寒冷的地方，你将在冰雪的包围中看热气袅袅上升，体验何谓真正的冰火两重天。

虽然丰富多彩的娱乐活动让客人根本不想睡觉，但客房仍旧值得一说。酒店的客房十分宽敞，即便是标准间也毫不吝啬对空间的使用，棕色厚绒地毯搭配褐色木制家具，让空间充满暖意，壁纸上的树皮纹路及窗帘上的枝叶图案则呼应了树林的主题。踏进洗浴间内，一股热气顿时从脚底袭来，原来这里安装了地暖系统，类似设施还有公共区域的电子壁炉，让人对酒店的供暖印象深刻，单从温度上来讲，你绝对感受不到正值北国的冬天。

翌日清晨，当你拉开客房的窗帘，一定会被眼前的画面所震撼。此时刺痛你眼睛的不仅仅是强烈的阳光，还有一片白茫茫的冰雪世界。透过巨大的玻璃窗，能看到蜿蜒的雪道和往来不息的缆车，顺着雪道及缆车的方向望去山顶，技高人胆大的滑雪者们正尖叫着俯冲而下，身后溅起一路雪粉，惊险如武侠片中的打斗场景。

Club Med Yabuli

As the first holiday village of the international holiday group Club Med in China, the Club Med Yabuli snow resort, designed by Nassau architectural design firm, provides a "one-price-all" services including accomodation, catering, transportation, entertainment and so on. The bustling entertainment and outdoor ice and snow scenery have let people experience the icy and hot situation in personnel.

At northeast part of Heilongjiang Province, the snow season has covered almost half of the year. Its winter is cold extremly. At four o'clock in the afternoon, Yabuli has enter its night time. In the dark, the resort hotel seems like a snow castle.

Entering the Club Med, you shall read this instruction first, or you will miss a lot of wonderful activities, such as dance course, yoga trainning, aerobics, table tennis, billiards, mahjong, trampoline, and a variety of settings for children entertainment. There is even trapeze circus course which you have never thought about. Of course, all these courses are free of charge. Therefore, we could find the sweat people taking their sports. In addition, this hotel also provide a a lively place--Forest Bar for nightlife. The bar's overall design is taking Wuhua Mountain located in Yabuli as the spirit. The furniture is adopted five colors of the mountain. Among them, there are set pillars covered with tree-skin-pattern wallpaper. The special made carpet patterns are from the Wuhua Mountain surface diagram, while the bar table is re-interpretation of traditional Chinese medicine cabinet. At night, the Forest Bar is a bustlign scene. Guests can taste all kinds of wines here for free, as well as play Karaoke, or dance with G.O. Of course, for many guests, they have to have a battle with cold in Yabuli. Then, the outdoor Jacuzzi is the best place to face the cold directly. You will watch the hot steam raising to the sky in the snow place, let people experience the icy and hot situation in personnel.

Though the colorful entertainment letting people have no sleepy sense, the room is need to introduce also. The room of this hotel is spacious. Even for the standard room, its space is not less any more. The brown thick velvet coupled with the brown wooden furniture, has made the space full of warm sense. The tree-skin-pattern wallpaper

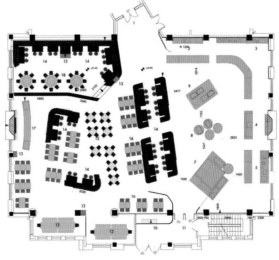

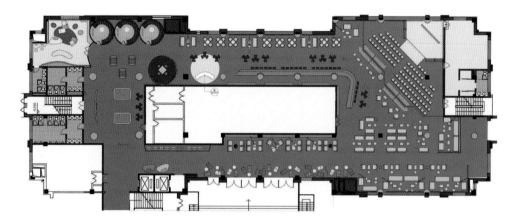

and the leaves pattern on curtains are echoing the theme of forest. Entering the bathroom, a stream of heat wave suddenly hit from the soles of the feet. Then, we find that there is installed a floor heating system, as well as the electronic fireplace in public areas. People will have a deep impress to this hotel. Just from the temperature point, you will never feel that the outdoor is in deep winter.

At the next morning, when you open the curtains, you must be catched by the scene. The shining sunlight and the icy and snowy world are sting your eyes. Through the huge glass window, you could see the winding snow road and coming&going cable car. Along with the snow road and cable car direction, you could find the skilled and daring slipping down. The snowpower is stirred behind, thrilling as the fight scenes of martial film.

REGION
NORTHWEST
CHINA

西
北

马柯艺术家餐厅

项目地点：西安市
设计单位：西安鼎马柯装饰设计工程有限公司
设计师：马安平
主要材料：不锈钢、金刚板、防古砖、玻璃钢
摄影师：申强

　　马柯艺术家餐厅所在地位于西安市的南郊,是有名的商业区,更是高校密集的大学区,这里分布着包括西安美院、西安音乐学院在内的众多高校,同时还毗邻雕塑一条街,可谓是闹中取静,艺术氛围浓厚。

　　马柯艺术家餐厅位于大厦负一楼,是一家以音乐、雕塑、油画、收藏、美食为一体的主题餐厅,主要经营经过改良的新概念泰国菜。餐厅总面积约为1000 m²,主要分为共享大堂、包厢以及办公区三个主要功能空间。

　　这里的餐厅老板同时也是一位画家,平时酷爱收藏,在"独乐"之余,也想借着餐厅来实现"众乐"的初衷。所以,面对这样新奇又独特的餐厅,第一次来的人,往往都对这样的"一览无余"印象颇为深刻。

　　大堂内的就餐位置不多,且间距很大,因此整个餐厅为食客营造了一份开阔、自由的就餐氛围。该空间在设计上引入"画廊"的理念,将餐厅四周以及隔断处都设置为画廊,内部陈列着的都是老板亲笔的得意之作。

　　除了大堂之外,餐厅还拥有四个以艺术家命名的10~30 m²的包厢。对应主题艺术家,包厢之间各有特色,有的简约精致、有的色彩浓烈。因由周边画廊的冷光的作用,以及墙面设计的镂花有机玻璃隔断,为就餐者提供了良好的私密性保障。

　　由于餐厅多了许多艺术品的展示,使得就客人在就餐的过程中,有了别样的新奇体验。

　　餐厅内的这些陈列品,有中国古代的旧门、清代洗手盆、明代圈椅、旧案几、闷户柜,有源自玛雅文化的雕塑,有解构主义的装置艺术,有荷兰现代家居品牌moooi中的马灯……甚至在入口处老门的背面还摆放了一台钢琴。从中国的到西方的,从古典的到现代的,在同一个空间内充斥了如此之多的风格和元素,却不给人以排斥和杂乱感。

Make Artist Restaurant

Make Artist Restaurant is located at the southern suburb of Xi'an, which is a famous business quarter. Here stand a large number of colleges and universities such as Xi'an Academy of Fine Arts and Xi'an Conservatory of Music. Besides, it is next to the Sculpture Street. Immersed in rich art atmosphere, the location is full of both liveliness and quietness.

Make Artist Restaurant is on the ground floor of the mansion. It is mainly engaged in improved new-concept Thai cuisine, with music, sculpture, painting, collection and cuisine as its theme. The restaurant totally covers an area of about 1000 m², divided into three main functional spaces, i.e. public hall, compartment, and office area.

The owner here is a painter who likes collecting very much. He enjoys this happiness and wants to share it with every one through the restaurant. So, such a novel and unique restaurant will impress people who come here for the first time with its "panoramic view".

The hall holds not too many seats, but space among them is large. In doing so, the restaurant provides customers with an open and free atmosphere. The idea of "gallery" is introduced into the space to design the around and partitions as galleries where the owner's favorite works by himself are displayed.

In addition to the hall, the restaurant has four 10-30m2 compartments named after artists. Corresponding to each artist, each compartment has its own features, some simple but exquisite and some colorful. Owing to the cold light of galleries around and engraved-glass partitions on wall, customers' privacy can be well kept.

The exhibition of art works in the restaurant offers customers new experience during dining.

The exhibit in the restaurant includes things of ancient China, such as old door, hand basin in Qing Dynasty, round-backed armchair in Ming Dynasty, old table and stuffy household cabinet; sculpture of Mayan civilization; installation art of Deconstruction; hurricane lamp of Dutch modern household brand moooi⋯.even the piano placed at the back of the old door at the entrance. In spite of so many styles and elements from China to the West and from the classical to the modern in the same space, people will not feel they are in disorder and exclusive to each other.

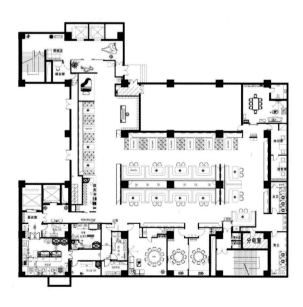
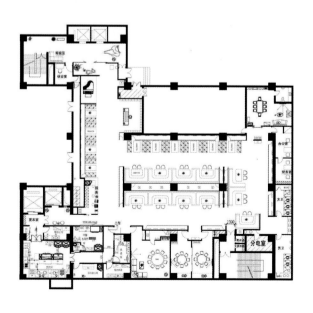

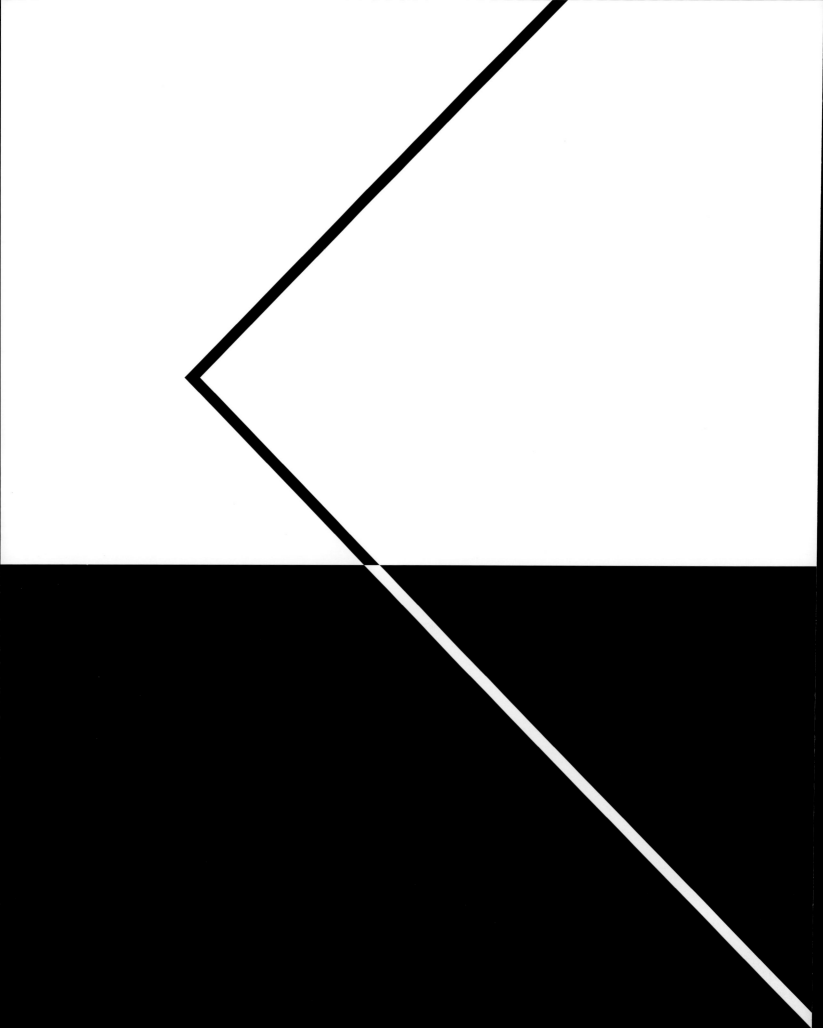

REGION SOUTHWEST CHINA

西南

拉萨瑞吉度假酒店

项目地点：拉萨
设计单位：Denniston International Architects & Planners Ltd.
设计师：Jean-Michel Gathy

被古代藏民奉为"诸神之地"的拉萨是西藏的精神与政治中心。坐落于古老的八廓街的拉萨瑞吉度假酒店位置便利，周围尽是各类文化地标，而宾客们则能够在这些充满传奇色彩的建筑物中探索拉萨的终极魅力。这座令人惊叹的度假酒店四周环绕着幽美的自然环境与拉萨那些震撼人心的山脉美景，酒店致力于为西藏带来更高品质的服务水准与奢华享受。富丽堂皇的度假酒店为宾客提供特色瑞吉管家服务、三家融汇各国美食的世界级餐厅、一间著名的品茗轩、一个酒吧以及面积达到 1087 m^2 的 Iridium Spa 水疗中心。这里的泳池装饰华丽，散发着金色的光芒。另外，宾客还能够在沉思园中锻炼瑜伽、进行普拉提训练或者静思冥想。

拉萨瑞吉度假酒店共有 150 间装饰豪华的客房及 12 套宽敞的别墅与套房，其中还包括总统套房。酒店将西藏的丰富文化及异域自然风情完美地融合在一起，从设计之初便考虑到了诸多可持续发展方案，包括太阳能电板、酒店餐厅选用来自本地的农产品与草本，并设有地下水循环系统。内部装饰精美的酒店客房配备了设计独特的家具、宽敞的大理石浴室、奢华床单以及诸如高速网络连接与等离子电视等设施。

瑞吉还将在拉萨瑞吉度假酒店首次推出最新的水疗品牌 Iridium Spa。旨在成为世界最高奢华级别水疗品牌的 Iridium Spa 能够令宾客彻底放松身心，重新焕发活力，其专为个人定制的水疗体验则包括传统西藏草本护理，以及使用本地滋养草本与植物（例如柏树叶及杜鹃花叶）的香薰理疗服务。Iridium Spa 水疗中心融合了五星奢华体验以及神秘的西藏精神疗法，这里的恒温泳池富丽堂皇，闪耀着金色光芒。另外，宾客还能够在沉思园中锻炼瑜伽、进行普拉提训练或者静思冥想。

拉萨瑞吉度假酒店的餐厅装饰极具文化气息，而这里的多种美食体验也会让人胃口大开。酒店的招牌餐厅秀（Social）拥有各类国际美食，全天候为您提供便利的餐饮服务；而斯自康则是第一家真正的藏式餐厅，这里环境优雅，令人倍感舒适，纯正的西藏美食与尼泊尔佳肴定会令你垂涎不已。宴庭（Yan Ting）共有六间私人包厢，宾客们能够在这里品尝各式各样的粤菜与川菜佳肴。

作为西藏首家奢华酒店，拉萨瑞吉度假酒店荣耀推出本地区第一家 Decanter by Haut-Brisson 酒吧，这里藏有140多种各类美酒以及陈年雪茄。品茗轩典藏了多种当地与进口名茶，而拥有10至25年茶龄的陈年好茶则更是令人陶醉。在西藏这也是第一次能够拥有这样的别致服务。

St. Regis Lhasa Resort

Lhasa is regarded as an spiritual and political center. Located in the old Barkhor Street in Lhasa, St. Regis Resort is conveniently located, surrounded by all kinds of cultural landmarks, where its guests can explore the ultimate charm of Lhasa among these legendary buildings. This stunning resort hotel is surrounded by beautiful natural environment and stirring mountain views of Lhasa, and it is committed to presenting higher quality of service standard and luxury enjoyment to Tibet. The magnificent resort hotel provides guests with the characteristic St. Regis butler service, three World-class restaurants integrating international food, one famous teahouse, one bar and one Iridium Spa in an area of 1087 m^2. The swimming pool here is beautifully decorated and exudes golden lights. In addition, guests can also practice yoga, Pilates or meditate in the meditation garden.

St. Regis Lhasa Resort totally has 150 luxuriously decorated rooms and 12 spacious villas and suites also including the Presidential Suite. Considered with many sustainable development programs at the very beginning, the design for this Hotel results in a perfect harmony between rich culture and exotic natural style of Tibet, for examples, the hotel has solar panels, offers the food available in restaurant of the hotel made from local farm products and herbs, and operates a groundwater circulation system. The hotel rooms with beautiful interior decoration are equipped with unique designed furniture, spacious marble bathrooms, luxurious bed linen, as well as facilities such as high speed internet access and plasma TVs, etc.

St. Regis will also initially launch Iridium Spa, the latest spa brand at the St. Regis Lhasa Resort. For the purpose to become the world's highest luxury level spa brand, Iridium Spa can help

guests to completely relax body and mind, re-energize vitality, whose spa experience designed for individual customization includes traditional Tibet herbal care as well as aromatherapy physical therapy service using local nourishing herbals and plants (such as cypress leaves and azalea leaves). Iridium Spa center combines five-star luxury experience and mysterious Tibetan spiritual therapy, where swimming pool at constant temperature is magnificent, shining with golden lights. In addition, guests can also practice yoga, Pilates or meditate in the meditation garden.

The decoration of St. Regis Lhasa Resort's restaurants is extremely full of cultural atmosphere, where a variety of culinary experiences also make a good appetite. Social, the hotel's classic restaurant, has all kinds of international cuisine, providing you with convenient food and beverage services around the clock; while Sizikang is the first pure Tibetan-style restaurant with elegant environment, where people feel more comfortable, and pure Tibetan food and Nepalese cuisine will make you salivate. Yan Ting has totally six private dining rooms, where guests can enjoy a wide range of Cantonese and Sichuan cuisine.

As Tibet's first luxury hotel, St. Regis Lhasa Resort Hotel has gloriously launched the first Decanter by Haut-Brisson bar in local collecting more than 140 kinds of wine as well as vintage cigars. Teahouse collects a variety of local and imported tea, where the good tea collected for 10-25 years is even more intoxicating. Tibet is also the first time to have such a unique service.

成都岷山饭店

项目地点：四川省成都市人民南路二段55号
设计单位：YAC（国际）杨邦胜酒店设计顾问公司
设计师：杨邦胜
建筑面积：25000 m²
主要材料：米黄石、金属、琉璃、玻璃、仿古砖
摄影师：贾方、马晓春

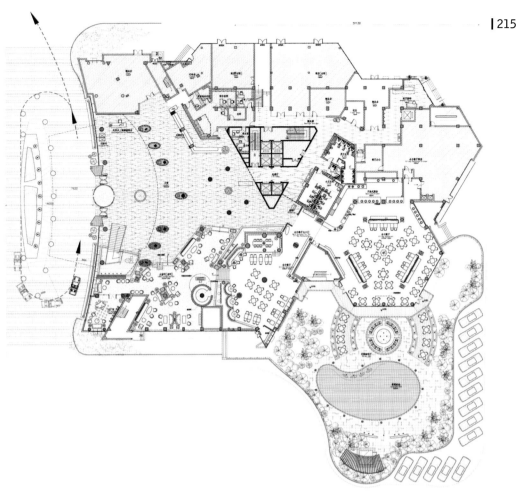

该作品为翻新改造设计项目。建于上世纪80年代的成都岷山饭店，是西南地区最早的高星级酒店，由于年代久远，亟需翻新改造。设计中，岷山饭店被重新定位为精品设计型城市商务酒店，设计紧扣川蜀地域文化特色，从川蜀文化中提取休闲文化、岷山文化等设计元素，将其化作酒店内上百只倒扣茶碗吊顶、黑白分明的毛笔灯饰群、随处可见的岷山水墨图案，成都的"市花"芙蓉也变成了大堂空中悬落的芙蓉花瓣灯饰，呈现一幅隽永的川蜀水墨画卷。作品运用国际化的设计理念和表达方式，使改造后的岷山饭店重现昔日的辉煌与荣耀。

This piece of work is a re-shaped design project. Built in the 1980s, Chengdu Minshan Restaurant is the earliest star hotel in southwest. Due to long history, then it need to be re-shaped once immediately. In the design, Minshan Restaurant is re-positioned as a boutique urban business hotel. This design is close to the local culture, such as leisure culture and Minshan culure extracted form Sichuan province culture. In addition, the designer has converted them into hundreds of inverted bowls on the ceiling. They black and white color lights, Minshan Mountain ink pictures can be seen easily. The municipal flower of Chengdu city Furong Flower has become a Flos Hibisci Mutabilis Light hanging in the lobby, showing a meaningful Sichuan ink volume. This piece of work adopts the international design concept and expression way and makes this re-shaped Minshan Restaurant show its former glory and gorgeous.

Chengdu Minshan Restaurant

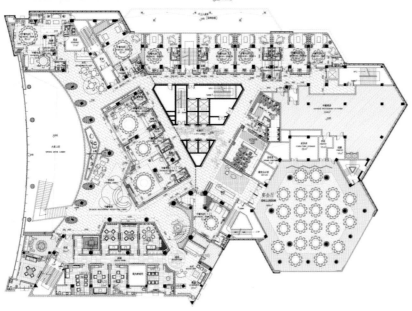

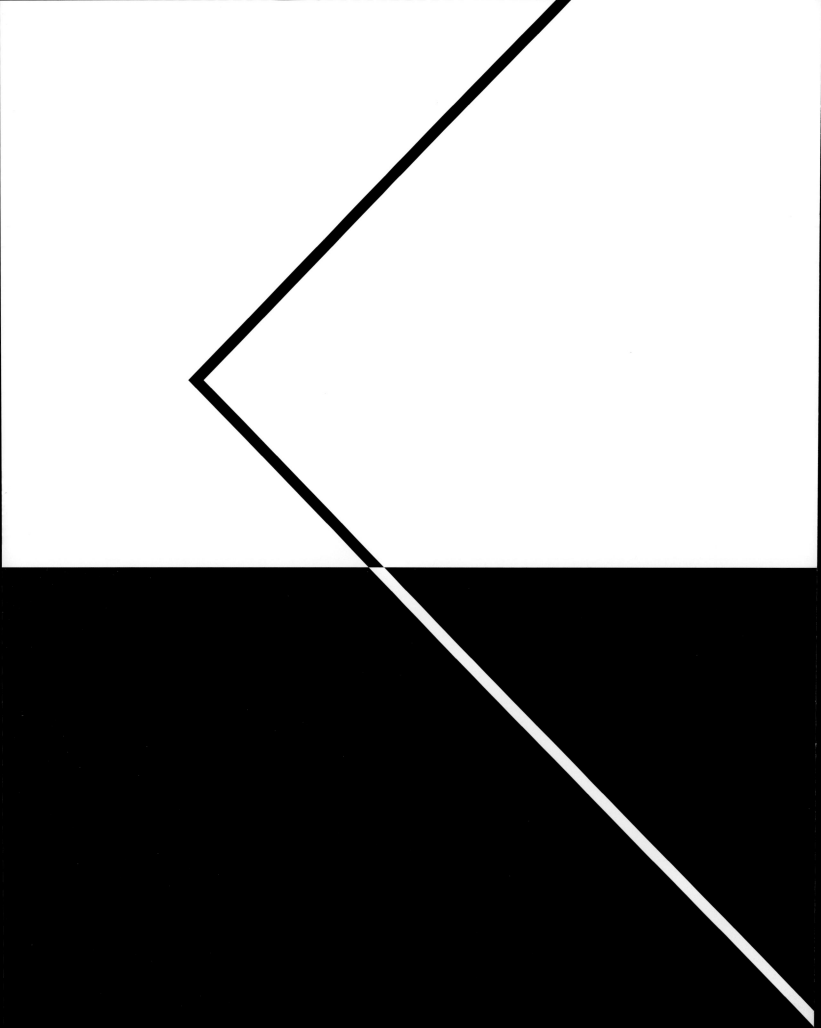

华东
REGION EASTERN CHINA

唐乾明月
福州接待会所

项目地点：福建省福州市置地广场7层
设计单位：道和设计机构
设计师：高雄
参与设计：高宪铭
项目时间：2011.05-2011.06
建筑面积：**350 m²**
主要材料：黑钛、墙纸、仿古砖、蒙古黑火烧板、绿可木、白色烤漆玻璃
摄影师：施凯、李玲玉

用现代手法演绎中式风格的新内涵,颠覆会所的传统概念定位,用标志性的中式元素赋予会所新的空间使命:宁静致远的环境,触手可及的高端。这就是以"新中式主义诗意栖居"为销售亮点的唐乾明月会所。没有醒目的介绍,看不到大张旗鼓的气势,藏身于市中心的五四路段,可谓是小隐于市。

初入会所时,见着空间上的明月造型和软装上相呼应的莲花灯时,脑海中猛的浮现出"明月耀清莲"的画面,不禁脱口而出:明月如霜,轻妆照水,纤裳玉立,飘飘似舞。

意境

别,通常理解为特别,植根于生活,跳脱于世俗,用轻描淡写的超然,直抵内心的柔软,刻画出的深邃印记,值得用忠诚追随。本案就是一个如此特别的地方。没有泛着冷光的大理石路面,没有高耸迷离的水晶灯,只是将向往的情境空降,让奢侈的精神之旅,在环境下,灯光里,水声中,不着痕迹地实现。

在会所的入口处随意树立着若干木桩,有着一种原生态的感动。中式元素的运用,行云流水,一如清莲般恰到好处的出现在属于它的位置上,流畅的视觉引导,自然不做作。只在你伫立或是冥想之余,一杯茶就可品得它的万般风情。灯光下,稍作修饰的原木,泛着淡黄色的光泽,一种温暖在蔓延,涌动着某种喜悦。直白地展示在眼前的木质纹理,显得格外的亲近,让人有一种想触摸的冲动,引得内心里回归自然的本能蠢蠢欲动。

走廊的尽头,一个旗袍形状的木架上,大红灯笼高高挂。红色,在原木色系的映衬下,少了艳丽之感,印在墙上的红晕,窥视出设计者心思的细腻:赋予配饰思维。用它们的形态,绽放设计光彩,延续设计生命。正如用潺潺的流水声,随意拨弄着你的情感起伏。

诗情

行走于会所内的各个角落,才发现空间格局划分上采用半封闭的隔断。不仅实现了视觉上的丰满,气氛上多了点互动的轻松。有违常规的细节调整,兼顾了会所的私密与社交。

习惯是那种浸入骨髓,自由地支配着你的行为和思想的方式,不自觉中就成为了生活的部分。会所里,半封闭的隔断造型就如习惯般如影随形。透过磨纱材质的玻璃,来往的人影宛若月朗星稀的夜空里,轻覆于皓月表面的那抹柔纱,显得空灵而悠远。我想,这才是设计的高明,用一种情境的标志,融合在空间的存在美感,造就身临其境的真实。似远非远,似近非近,切身感受着想象的美好,貌似遥不可及,一个转身却又在灯火阑珊处。夹杂在其间的若干绿植陪衬,随人影走动带起的花瓣摇摆,轻灵得可爱。掀开垂帘,圈椅上的年轮清晰可见。阳光下的尘埃,就是那似是而非的沉淀。清冷内敛的莲花手座,将禅化解,散落的,是对生活的悟。

离开时,再次浏览了墙上的《爱莲说》。设计何尝不是在百花争艳中的市场中,用它的格调对味着不同的有识之士。好的设计,会说话。它出淤泥而不染,濯清涟而不妖,即可远观也可亵玩。

Tangqian Moon Fuzhou Reception Club

It adopts modern style to perform Chinese style, toppled down all the traditional positioning concept, taking the symbolic Chinese elements on the new space: that is quiet but meaningful environment, touchable highness. This is Tangqian Moon Club that takes "new Chinese poetic dwelling" as its eye-catching point. There is no brilliant introduction, no much exaggerated momentum. It is hidden in the Wusi Road of urban center, which can be described as humble one of the city.

The moment you seeing the club, the moon shape of the space and the lotus light lamp will give you a feeling that "Moon smiles to the nice lotus", which make us remember this sentence: bright moon as snow, nice beauty stand there, like flying dance.

Meaningful Environment

Different, means particular in general. Taking life as the basic, escaping from the world, it adopts simple decoration to touch our soft soul and leaves deep impression to us, which is worth pursuing with our honest heart. Tangqian Moon Club is such a special place. There is no marble pavement with cool light, no towering crystal lamps. There is only dreamful environment, letting the luxury spirit tour realized under such environment, light, and ripple of stream.

At the entrance of this club, there are some wooden pegs, which give us a natural feeling. The adopting of the Chinese elements is such naturally, just as lotus is planted at the proper place, eye-catching and elegant. When you stand here or meditation, a cup of tea will bring you warm feeling. Under the light, the slightly modified wooden pegs, there is glowing yellow light, as if seeding a kind of warm, full of joyous. The original wood texture is facing us directly, which looks particularly familiar with, letting people want to touch it to feel the real natural beauty.

At the end of this corridor, a cheongsam-like wooden frame, several red lanterns are suspending there. Red, under the original wood color system, is lack of some gorgeous. Its flush is reflected on the wall, showing the designer's exquisite mind: that is given mind to the accessories. Taking their morphology, let the design brilliance and extent his design life. Just as the gurgling sound, they are touching your soul before you realize it.

Poetry

Walking among each corner of the club, we find that the space is adopted semi-separation way. This has not only realized the fullness of visual, but also relaxed the overall atmosphere. The conventional details has both considered the private and social aspects.

Habits are one that immerse our bone and dispose our behaviors and thought freely. It has become a part of our life unconsciously. In this club, the semi-separation structure just as the habits.

Through the glass, the people that come and go are like the starry sky, covering on the surface of the glass, seemingly much ethereal and meaningful. In my opinion, that is the good point of this design. Merge a kind of mood into the space to show the reality of scenery. No matter far or near, you can feel the imaginary beauty. It is seemingly out of reach. But the imaginary beauty is just at the dim light part. Several green plants are scattering there, flowing with the comes and goes, smart and lovely. Open the curtain, the armchair on the ring is visible clearly. Dust in the sun is the paradoxical precipitation. The humble lotus-like seat is interpreting the Buddha and the meaning of life.

When leaving, I read "On Lotus" on the wall once again. This design is just for the talents people. A good design can speak itself. It will clean and elegant even come from mud. It will simple and humble even with beautiful face. It can be watched both in far and near distance.

熏茗茶叶会所 湖东店

项目地点:福州市湖东路
设计单位:福州北岸设计有限公司
设计师:王家飞
建筑面积:800 m²
主要材料:楼兰仿古木地板、硅藻泥、青石
摄影师:周跃东

本空间为一处高端茶叶会所。设计师以江南小镇作为设计的灵感来源,以室内剧的艺术手法,融合现代的美学价值来诉说以个小镇的故事。窄小的街巷、昏暗的路灯以及小桥流水人家,小镇的生活气息扑面而来。同时,设计师借助于这样的意境表达了对中国传统文化的尊重与敬仰。

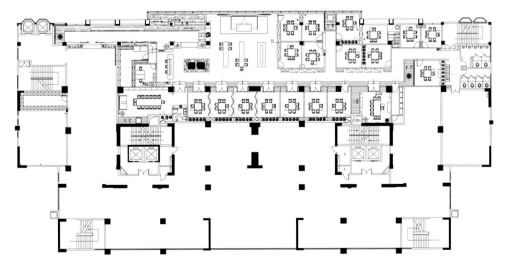

Ximing Tea Club Hudong Shop

This case is a high class tea club. The designer is taking the small town of South Yangzi River as the inspiration, taking the artistic techniques, merging with the modern aesthetic value to tell the story of a small town. Narrow streets, dim lights and bridges, houses, the town's life has run to you. At the same time, the designer has expressed the respect and admiration to Chinese traditional culture in this way.

一信（福建）投资办公室

项目地点：福建三明
设计单位：福建国广一叶建筑装饰设计工程有限公司
设计师：何华武
建筑面积：**1800 m²**
主要材料：玻璃纤维、u型玻璃、仿古地砖、生态木、灰木纹大理石、地毯、玻璃、张拉膜
项目时间：**2011.02**
摄影师：周跃东

本案是一家房地产开发投资公司的办公室，面积约 2000 m²，由于该公司业务涉及建筑与房地产，因而办公室的设计以体现强烈的建筑感为主。在人、环境、建筑、空间之中寻找出设计主线，彰显自然的建筑美感。跃层式的建筑在环绕着中庭的楼梯中，透过人与空间的互动因子，拉近生活与工作之间的距离。活力、创新、时尚、独特是公司始终坚持的理念。这在前台与公共区域的设计上均有体现，借由一个具有强烈造型感的前台与折线的中庭楼梯巧妙的结合，在视觉上形成一个由点、线、面组合而成的多维几何空间。整体空间充满几何造型的美感。加上前台柜子的序列导向感与楼梯的不同形状塑造出折型的力度，展现出一种健康与积极的空间语言，更有独特、时尚、创新的空间视觉延伸感。让人们一走进大堂立刻能感受到积极向上的企业文化。

整体空间主要是以黑白灰作为主轴，纯净视觉及心灵，将外界的纷扰逐渐沉淀，让工作回归于宁静，给予空间价值。成就简洁、时尚、自然，使内部与外部的空间在视觉上达到高度融合。我们所做的不仅让它变得更美丽，而是为了彰显进取精神。

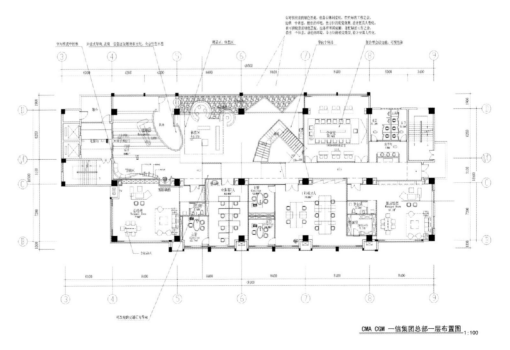

CMA CGM 一信集团总部一层布置图 1:100

Yixin (Fujian) Investment Office

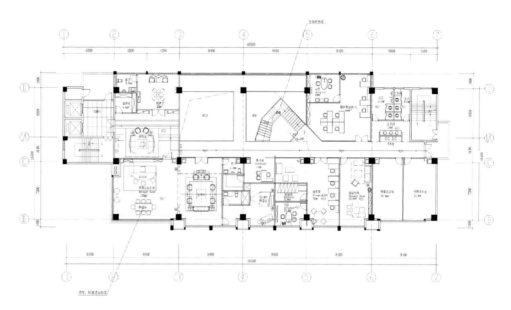

CMA CGM 一信集团总部二层布置图 1:100

This case is an office invested by an estate investment and development company, covering 2000 square meters. As this company refers constructure and estate, the office design shows a strong construction sense. Among people, environment, space, they choose the design theme, showing the construction sense naturally. The jump style construction has shorten the distance between life and work through the spining stair. They have formed a multi-demensions through combining point, line and face together. It is expressed in both reception table and public area in design. By this strong shaping sense,The whole space is fulfill with geometric modeling beauty. Coupled with the re-

ception table as well as the different shape, they are showing a healthy and active space language. There is even unique, fashion and innovation space as the extended scenery. You will feel the actively enterprise culture the moment you enter.

The whole space is mainly taking black and white as the main tone to purify visual and soul, precipitating the hustl and bustl and let the work be quiet, making the space more valuable. Creating simply, fashion and nature, make the internal and external space combined together in a highly degree. What we have done is not to make it more beautiful but to show the progressive spirit. At present, the design industry talks more about the positioning of landmark style. We believe that, like out team can not only limitein the landmark style, we shall have proper response to the environment. At the beginning of each project, we all shall take it as the improvement of us.

位于福州西湖湖畔的静茶吸取了西湖凝聚千年的灵气，在周遭一片灯红酒绿间突显着它的静谧和高雅。设计师沿用了"禅茶一味"的设计理念，将东方传统文化中"禅"的思维方式来体现繁华落尽的细微之美，将静茶的精髓完美呈现，同时成就空间的空灵之美。

禅是东方古老文化理论精髓之一，茶亦是中国传统文化的组成部分，品茶悟禅自古有之。以禅的风韵来诠释室内设计，他不求华丽，旨在体现人与自然的沟通，以求为现代人营造一片灵魂的栖息之地。会所内以素色为主调的痕迹，粗糙的青石板与天然纹理的仿古砖厚实而流畅，仿佛划过满了时间的痕迹，为整个空间带来一种大气磅礴的气势。而被大面积运用的源于中国古代窗花造型演变而来各式花格，则以一种独特的姿态诠释着中式之美。为了柔和这些硬朗的材质，设计师在细节之处为用心，无论是泛着暖昧红光的伞形灯、走道上看似随意摆放的佛像、枯枝，还是那些做工精良的中式家具，置于展示柜内的精美瓷器与茶具，这些细微之处的累积都让空间显得更为饱满。

光是内部空间里不可或缺的角色，本案利用光的明、暗、虚、实等属性给室内带来充满变幻的视觉享受。室内一暖色射灯作为空间的主要装饰光源，暖而不媚。利用格栅镂空使光线自由地在室内游走，随性洒脱。此外，设计运用墙面、柱子、玻璃、隔栅等作为空间的视觉间隔，使得其中的相关区域或灰暗狭长或宽敞明亮，戏剧性的明暗对比犹如一曲抑扬顿挫的咏叹调，令其回味无穷。

项目地点：福州
设计师：施旭东
设计单位：福州旭日东升装饰机构
建筑面积：400 m²
主要材料：黑色方管、爵士白大理石、黑镜、仿古砖
摄影师：周跃东

静茶西湖店

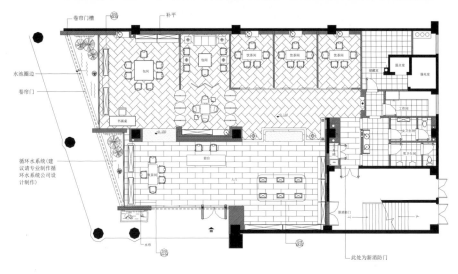

Taste of Tea and Buddha

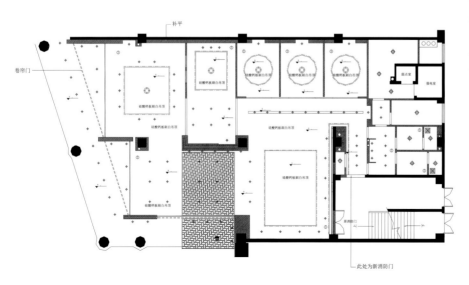

Jingcha, located at the bank of Xihu Lake, absorbs the spirit and soul of Xihu. It looks more quiet and elegant among the bustle and hustle surroundings. The designer adopts "Buddha and Tea" design concept, which has make the traditional oriental culture "Buddha" thinking way to show the beauty of the city. The soul of Jingcha has been fully shown, making it more charming.

Buddha is one of the essences of oriental ancient culture. Tea is also part of Chinese traditional culture. Drinking tea and meditating Buddha have been existed for a long history. Taking Buddha's elegant to demonstrate the indoor design, it does not pursuit of luxury but mainly to show the communication between human and nature as well as create a pure soul and heart space for modern society. In this club, the major tone is light color. Looking at the rough bluestone that combined with the natural texture of antique brick, smooth and thickness, they seem like that have possessed time traces and brought the space with majestic beauty. In order to warm these hard materials, the designer has paid much attention on the details, no matter of the warm red umbrella-like lamp, or the Buddha statues and dead branches placed on the road naturally, even these well-crafted Chinese furniture, as well as the fine porcelain and tea cups. These accumulations of details have made the space more colorful.

Light is an unnecessary role of the interior space. Jingcha uses light's bright, dark, virtual and real nature to bring the changing visual enjoyment for space. The warm spotlight of room is the main decoration light source, warm but charming. The hollow grid allows the light can walk in the room freedom and easily. In addition, using the wall, pillars, glass, grid as the separated things makes the relevant area be dark, bright, long or gray. Such changeable light likes a poem with ups and downs tone, letting people immerse into it.

天瑞酒庄

项目地点：福州
设计单位：宽北设计机构
设计师：施传峰、许娜
建筑面积：**180 m²**
主要材料：瑞士卢森地板、墙纸、大普家具、万事达灯具、西门子开关、荔枝面大理石、软木等
项目时间：**2011**
摄影师：周跃东
撰文：江雍箫

斟一杯法国红酒的浓香，赏一抹意大利红酒的色泽，品一口葡萄牙红酒的甘美，在浓浓诗情画意的中品尝真纯佳酿，浪漫尊贵随之弥漫。在天瑞酒庄里，每一个转角几乎可以视为对葡萄酒文化的传承与演绎。它的空间情趣与节奏风格融合了多样的风情与文化，使得隐藏于都市人心中关于精致生活的那些奢望落到了实处。

酒庄共被分为上下两层，它们之间彼此独立，却又不乏交流的可能。空间中的不同区域在满足各自功能的基础上，用色彩、光影、材质的变化来引导着人们的视觉享受。这里似乎在改变着我们对时尚装修概念的定性思维，设计师充分利用了与红酒相关的元素尽情演绎了多元的红酒文化。一楼入口的锥形柱做成由夸张变形的大"橡木塞"重叠而成的形状，它既是纯粹的装饰片段，又是一种时尚的演绎；洗手台旁放置着装饰品与酒杯的"高几"，竟是一个古朴的橡木酒桶，它毫不隐晦地表现着其率真与坦诚的面孔；而用软木塞串成的帘子则成为了一大面积的背景墙，为我们带来了新鲜的视觉体验。当射光打在上面时，仿若满墙的灿烂繁星，远观着则又像折射着光的瀑布，似乎当一阵风吹过，能看到晃动的帘子后藏着若隐若现的宝藏。

除了材料上的新奇，空间里的色彩与灯光设计也控制着来访者的心情。置身其中，会有一种奇妙的感觉，仿佛从现实的喧闹中走出来，而后在这个暖色调的氛围里渐渐褪去那份浮躁。设计师匠心独运地将点光源与泛光源进行有机地组合，并用独特的灯具造型来丰富空间的美感。当似虚而实的光影透过栅格、屏风铺洒在周遭，影影绰绰地构筑起了一方新奇的空间，仿若在梦中。而随着人们脚步的临近，飘渺的梦境也一点点清晰起来。因为视角的不同而产生出这种不确定的美感，使得灯光在赋予空间柔和的特质之外，还营造出些许神秘的效果。这是设计风格上的一种变调，亦可以是设计语言中一种出乎意料的洗练，让人们发现这里的每一个层次皆有动人之处。

在这个纯粹的空间里，或品酒或交谈，一切仿佛陌生，又好像分外熟悉。眼前的一切是如此得鲜活和可爱，而我们能做的只是运用辞藻作愉快的记录，并还原真实的场景。我们欣喜的是，面对这样的一个空间时，除了留下图文的记忆，内心竟是满足的。

Xiamen Guanghezuoyong

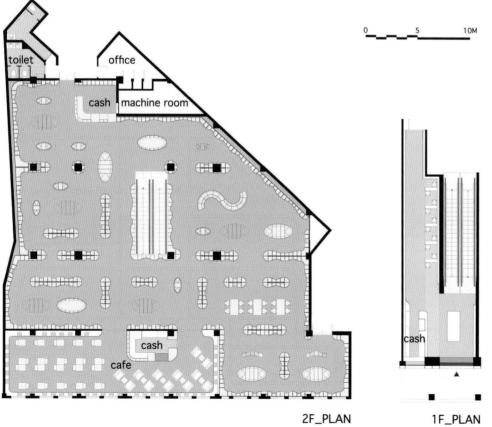

2F_PLAN   1F_PLAN

The design concept comes from the name itself "Photosynthesis". The "sky bookshelf" and "ocean bookshelf" that are hanging on the ceiling, as well as the "grain bookshelf" floating around the wall, and the "forest bookshelf" surrounding the pillars. Coupled with the green plants, these organic "landscape" has made the entire space full of life sense. Between the book shop and customer, there is generating the "photosynthesis" between knowledge and oxygen.

天墅销售中心

项目地点：厦门
设计师：凌子达
建筑面积：**600 m²**
主要材料：氟碳金属漆、橡木染色、玻璃、鹅卵石
摄影师：周耀东

这个设计案我们是从建筑到室内以及景观三方面同时整体构思完成的。本案为一个售楼中心，而案名为"天墅"，所以，以此案名为出发点构思概念。

而且项目地点位于厦门市中心，基地周围条件不理想，有旧公寓以及工地现场，正对面是一所学校，四周围并无景观可言。

最终是以"漂浮 Float"为概念，把整个售楼中心拉至2层的一个高度，并自己创造景观，一个叠级的水池，使售楼中心好似漂浮在水面上，并且来访参观者是走过大面水池，并穿过售楼处底部，走道后面的楼梯再上到售楼大厅。建筑结构为钢结构，建筑设计与室内设计力求手法的整体与统一性。

华东 REGION EASTERN CHINA | 261

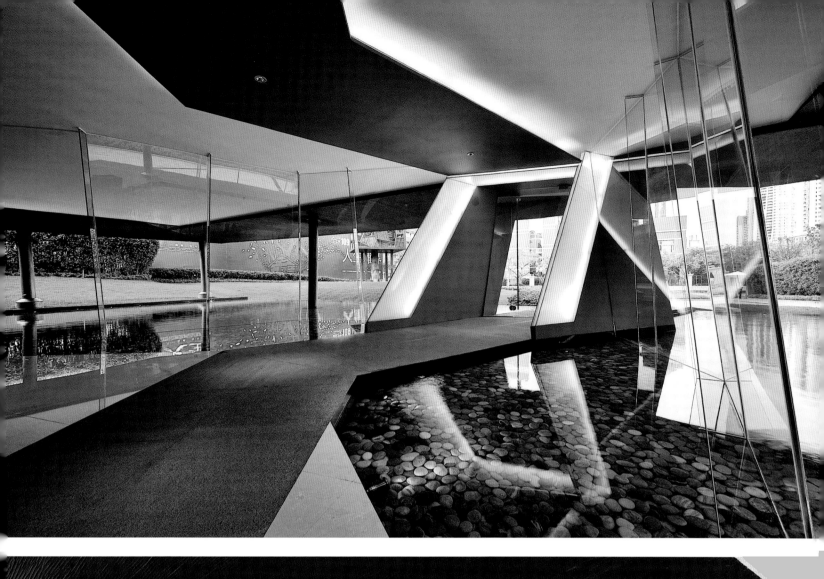

Tianshu Sales Center

This design case is finished by us from the consideration of construction, indoor as well as landscape. This case is a sales center, named "Tianye". Then we take this name as the starting point.

However, the project case is located at center of Xiamen city. There surrounding conditions is not that good. There are old apartments and project site, facing a school and no landscape.

Finally, taking "Float" as concept, we put the entire sales center in a same height with 2F and create landscape by ourselves. A multilevel pool seemingly has let the sales center floating on the water and allows visitors enter a water world. Passing through the first floor of the sales office, you will come on the sales hall by elevator. The contracture is steel structure. We seek to keep uniform in terms of the architecture design and indoor design

　　本案是多层的别墅，设计师摒弃了以往方正的入户空间格局，磨去了生硬的棱角，删繁就简，采用曲线造型巧妙地将入户空间延伸，使得视野变得开阔、舒展，原木材质的自然肌理配合简洁而柔和的线条，让平淡的墙面展现出沉静而舒展的气度，体现了一种"大美无言，大象无形"的中国士大夫的文化追求。室内外空间相互交融穿透，空间层次和材质整洁而洗练，给人带来朴实、安逸、闲适的自然体验，在一定程度上符合阴阳平衡、气场圆通的传统理论。

　　在客厅的设计上，简洁却不乏时尚的现代沙发、金属质感的半圆灯罩与原木电视墙和平共处。餐厅里层次分明，错落有致，简练的雕花柜门和生机勃勃的盆景，均赋予了现代空间独特的文化表情。

　　主卧的错层地台，巧妙地为宽敞的室内空间增添了趣味，体现了细节设计上的用心。云朵造型的床头灯及灯具的雨滴设计，兼具实用与审美价值，赋予静谧的休憩空间一份灵动的同时，也顺了中国传统天人观念中对于风调雨顺、阖家圆满的真诚祈福。

　　本案中，设计师并没有采用常见传统的元素如花格、书法等的表象符号，而是将中国传统的精神融入其中，使得作品极富自然和谐、顺达圆满且内敛含蓄的中国韵味。在这个快节奏的现代生活中，当人们想要摆脱那些不由自主的焦虑和恐惧时，我们不得不承认，那种古老传统的文化力量总有着如此那么神秘而强大的力量，给我们带来宁静与平和。

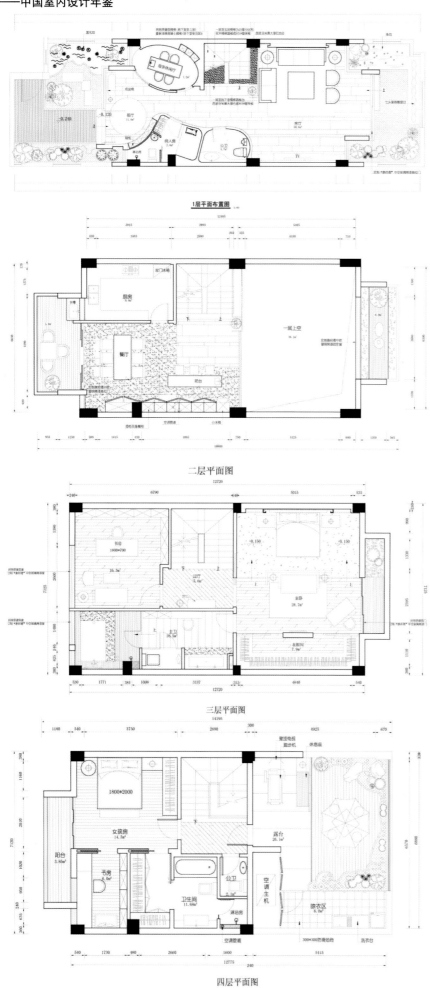

1层平面布置图

二层平面图

三层平面图

四层平面图

Distinguished Villa

    This case is a multi-floor villa. The designer has abandoned the traditional square framework, dropped the blunt edges, departed the complex and adopted simple. He adopts curved shape to extend this space clevely, letting the vision become open and comfortable. The natural texture of the wood is coupled with the simple and soft lines, which has let the normal wall show a quiet and comfortable mood, showing a Chinese literati culture pursuit of "Great beauty needs no language and no fixed shape". The internal and external space are interfaced with each other. The space structure and materials are neat and great, which would brought people simple, comfortable and quiet natural experience. To some extent, they have balanced the vitality and essence and the completed field in traditional theory. For the living room design, it is simple but fashion, such as modern sofa, metal texture semi-circular shape lamp and the wooden TV wall. The restaurant structure is reasonable. The concise carved doors and vibrant bonsai are given a unique cultural expression to the modern space.

    The split-level floor of major bedroom has added much funny to this space, showing the heart of designer. The cloud-like bedside lamp and the raindrop-like design bear both practical and aesthetic values. On satisfying the quiet sense to the bedroom, it also has conformed to the traditional Chinese concept of good weather and good harvest blessing.

    In the overall case. the designer do not adopt the normal traditional elements, such as flower pattern, calligraphy, etc. He merges traditional spirit into it, letting the works full of natural and harmonious, as well as humble sense. In this fast-paced modern life, when people want to get rid of the anxiety and the fear, we have to admit that such ancient traditional cultural forces is powerful, which will bring us tranquility and peace.

厦门海峡国际社区原石滩SPA会所

项目地点：福建省厦门市
设计单位：厦门喜玛拉雅设计装修有限公司
设计师：胡若愚
建筑面积：3000 m²
主要材料：贝芝文化石、威盛亚编织木皮防火板、LD瓷砖
摄影师：申强

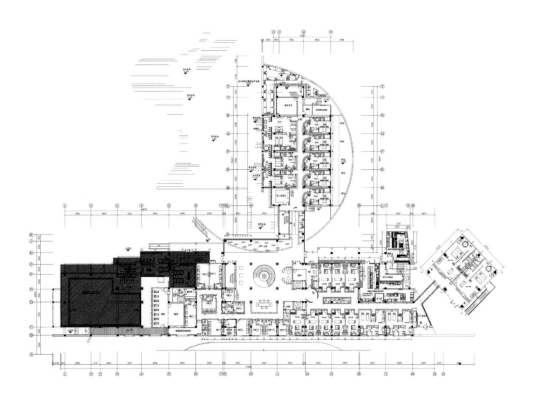

会所定位为高端品味人士量身打造的一放松身心的"桃花源"、躲避风雨的"避风港"。设计上既追求自然生态，应用原木、原石等自然生态材料，营造舒适轻松氛围，而局部又搭配红铜、皮草等材质，再加上精致的细节处理，彰显内敛奢华。风格上在简约的现代构图中，隐约着东方传统的雅气和禅意。空间布局上在公共部分或通过不断变化聚焦点让空间迂回曲折，或通过放大空间，用距离感来营造私密性。而在包间内部采用岛式布局，产生多回路的灵活变化、营造随性，无拘束的空间感受。设计及选材亮点：由结构柱上的石皮缝隙中，水流缓缓淌下，干湿浓淡间隐约着水墨意境；而不规则排列圆木倒映于水镜之上，池底星灯摇曳，营造气氛同时又遮蔽包厢间的视线；接待台后则是规则阵列的圆木，反射在天花的灰镜上，更显从容大气。接待台面采用整长厚实木料，浮于底部透光木纹石的光影之上。圆形的红铜管高低错落如芦苇一般，与红铜螺旋楼梯相映成趣。而贝芝石仿古面和木饰面的编织肌理，与整体风格相契合。

Original Rocky Beach SPA Club of Xiamen Channel International Community

This club is positioned at the high-class people to create a fully relexation "Shangari-la" as a harbour. This design is pursuing natural environment, adopting original wood, stone and other natural ecology materials to create a comfortable atmosphere.In the layout, it is coordinate with red copper, leather and other materials, as welll as some delicate treatment in details. They are showing a humble luxury. Its style is showing the traditional elegant and Buddist in the modern simple structure. Through winding on the space public part or changable focus, or enlarge the space, they are showing the private sense by distance. However, it is adopted island-layout in the room, letting people have a flexible, freedom and unconstraint feeling. The design and selecting materials' excellent point is that the water flow down from the gap between the stone, which is showing the ink-wash drawing sense. While, the irregular round wood reflecting on the mirror, the star-tyle light shining at the bottom of pond, they all create a better private space. Behind the reception table, there are regular round wood, reflecting on the gray mirror of the ceiling, showing grandness. The reception table is adopted whole and long and thick wood, floating on the shining stone. The round red copper pipes have different sizes, coordinating with the red copper stair in a funny way. While Beizhi Stone's antique texture is just in line with the whole design style.

巴厘香墅黄府

项目地点：厦门
设计单位：东方铭冠设计空间机构
设计师：曾冠伟
建筑面积：**500 m²**
主要材料：白玉石、黑板岩、黑白马赛克、
　　　　　强化木地板
摄影师：吴永长

这是一幢现代巴厘风格的别墅，含地下共4层，有坡屋顶，主人是一对80后的年轻精英，喜欢时尚、摩登的现代风格。通过对整体空间的研究发现，别墅后部三层的主楼与前部两层的裙楼之间宽1.5 m、长4 m的小天井很不实用，因为它的分割让一层的客厅与餐厅都显得狭小与逼仄，考虑到地下层还有前厅玄关至梯间及后院两个天井，保障了足够的透光通气，设计师决定将这个小天井从地下层到二层顶板逐层封闭，经过这个改造，每层的空间功能都变得顺畅而灵动。地下层原天井位置变成一个1500×3000（mm）的观赏鱼池，隔开了休闲影院与鉴品茶室，动静皆宜；一层天井位置成为室内空间，设置一岛式电视水吧台，既分隔出客、餐厅的功能区域，视觉上又融为一体且满足了区域的功能需求，整体空间一气呵成，变得宽敞明亮，空气的流通也更好；二层的天井位置变成一个半开放的室内小花园，紧临书房，凭添自然的气息；三层天井位置让主卧房前移1.5 m，直接与天台庭院相连，而床位前移后余出的空间正好安排一个梯间，将超高的坡屋顶利用起来，现浇一层楼板成为阁楼式的衣帽间，玻璃结构实木踏板的楼梯既是功能亦是炫酷的空间雕塑。至此，空间区域整理清楚，没有多余的造型，干净纯粹。

空间整好后，表面的装饰只需顺势而为就行。这时候色彩与肌理的选择将决定最终效果的成败。设计师用温润的白玉石，沉稳大气的黑板岩，时尚摩登的黑白马赛克及一块延续所有空间的温暖木色完成了现代、简洁、温馨、纯粹的空间表演，简单的四种主材演绎出黑白调永不过时的经典。客厅里红色渐变的沙发与窗外郁郁葱葱的绿意遥相呼应，是空间里最精彩的点缀。

黄府地下一层平面

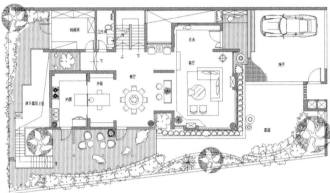

黄府一层平面

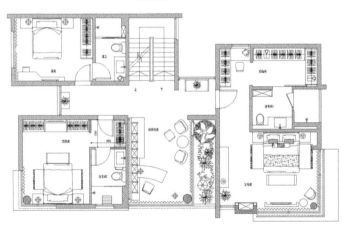

黄府二层平面

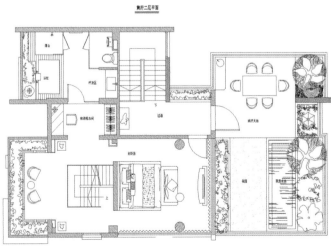

黄府三层平面

　　室内庭院的面积不大，无法进行真正意义上的修园造景。针对狭长的地形，设计师画了三道弧线就完成了庭院的规划，第一道内凹的弧线是 1.2 m 高的矮墙，墙后是 7、8 m 高的香樟树及竹子，墙前是灌木鲜花与草坪，形成入口前庭的端景；第二道外凸的弧线是 0.5 m 高的花池，正对客厅，巴厘风情的热带植物是客厅视觉的美丽延伸；第三道内凹的弧线又是一道 1.8 m 高的中墙，墙后还是 7、8 m 高的翠竹，墙体上有爬藤，墙前设置原木平台，是可以纳凉小憩的所在。在这个庭院的规划中，设计师将绿化树木做为边界背景来使用，所有的绿树翠竹都要求高过围墙，从视觉上将围墙消弥掉，营造"林"的感觉，意图产生让建筑融入自然的效果。所有的阳台、天台也都设置花池种上竹子，基本达到开窗见绿。在亚热带气候的厦门，这样的绿化屏障也最大限度的提高了隔热降温的功能，减少机械制冷的能耗，而经过树木过滤的清新空气对应了开敞的室内空间规划，让自然、环保、低碳的生活成为可能。

本案打破了空间本身过于规整的形态。设计团队以其独特的理解，将"设计、文化、艺术、生活"融入于空间当中，力求打造一座国际性的美学空间。映入眼帘的斑驳水泥结构与黑白块体的交织，与展馆中具有美感的、经典而华丽的家具形成了强烈的对比，让展品在视觉的碰撞下更为夺目。

CASA BELLA Home Art Center

Case, space itself is too neat to break the form. With its unique understanding of the design team, will "design, culture, art, life" into them in the space, and strive to create an international aesthetic space. Greeted mottled with black and white cement block structure of interwoven, and the exhibition hall with beauty, classic and beautiful furniture in sharp contrast to the collisions in the visual exhibits a more eye-catching.

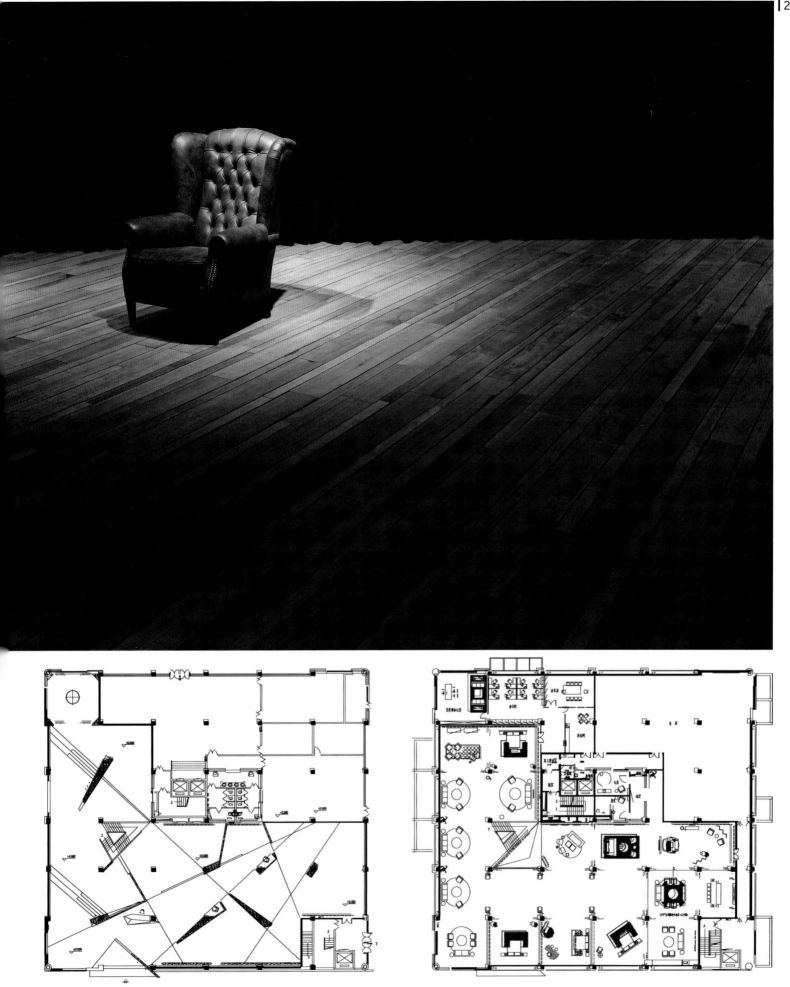

项目地点：合肥太湖路与马鞍山路交叉口
设计单位：北京法惟思设计
设计师：蔡宗志
参与设计：詹玉宝、童孝友
项目时间：**2011.10**
主要材料：木材、方管、石材、水泥
摄影师：孙翔宇

合肥卡伦比咖啡连锁

在进入设计之前，研究了关于咖啡的种种，在了解咖啡生产的地区，咖啡的品种，咖啡的烘焙等等中，渐渐的一幅咖啡地图浮现在眼前，有趣的是咖啡大部分产地集中在赤道附近，不同的海拔高度也生产不同的品种，如牙买加的蓝山咖啡在海拔1000 m以上，品种不同的咖啡豆也是有酸甜苦不同的分别，最后烘焙的方式不同也会影响咖啡的味道，学问之深，值得细细的品味。

忆起走在意大利及法国的任何街头上，常常会被一种蒸汽声伴随着咖啡香所吸引，入内，一阵吵杂声夹带着勺与盘的碰撞声，烟雾缭绕，一杯咖啡expresso，马上能体验到当地人的生活与咖啡文化的关系，如果再点个甜点，那真是完美的结合。在设计这个咖啡厅时，业主也能感受到我所描述的那种氛围。以这种氛围结合合肥当地的消费模式，提出了以咖啡操作台作为咖啡厅的视觉焦点，让每个客人在进入咖啡厅的同时，也能感受到，闻得到，不论是意式咖啡的蒸汽或虹吸式咖啡的慢火，让煮咖啡的人与喝咖啡的人之间产生一种交流，它是互动的。

在设计上，以咖啡操作台为中心，尝试把咖啡地图与咖啡厅的空间序列结合成为咖啡厅的一种新风格，用一组抽象的经纬度数字关系来诠释实际距离感与空间的高低差，让咖啡厅与咖啡地图串起来，以合肥店为北纬31度东经116度（31N116E），牙买加是北纬17度西经116度（17N76W），依索皮亚是咖啡的发源地1S36E，叶门产摩卡豆12N41E，巴西是世界第一咖啡生产国35W-74W，5S-35S，印尼是曼特宁咖啡产地等等，不同的数字除了代表不同的产地外同时作为包间的名称，心理上也缩短了实际间的距离，在空间的序列上有高低差是3D的关系及充满经纬的分割。

在空间轴上，咖啡操作台的下一个空间，设置一个书的空间（文化），内置有跟咖啡有关的书籍，也提供一些流行，设计等等的讯息，作为咖啡文化品味与知识的印证，增加咖啡的深度。

Hefei Kalunbi Coffee Chains

Before this design, the designer has studied many kinds of coffee. During study the coffee production area, coffees kind, coffee's baken ways and so on, a piccc of coffee map is appeared gradually. The funny thing is that most coffee production areas are focusing around the equator. Different height produces different coffee. For example, Jamaica Blue Mountain coffee is planted on the above 1000 meters high. Different coffee has different coffee beans. The baking ways also affect the taste of coffee. The deep knowledge is worth a great enjoy.

Remembering walking on the street in Italy and France, we are often attracted by the smell of coffee. Enter it, you will hear a burst of collision sound of spoon and plaste, smoke-filled all round. That's a perfect combination. When design this coffee shop, the owner also can feel the atmosphere i described. Combined this atmosphere with the local consumption patterns of Hefei, the design propose to take coffee making table as the visual focus of this coffee shop. It is mainly to let each guest join this coffee shop and feel and smell the Italy coffee steam or espresso or siphon coffee on the little fire. It is also a kind of communication between the coffee maker and gueas. It is interactive.

In design, taking the coffee making table as the center, the designer tries to combine coffee map and coffee shop's space together to form a new style. Adopting a group of abstract latitude and longitude figures to interpretate the gap between the actual distance and the space, merging the coffee shop and coffee map into a single whole. That is, 31N116E of Hefei, 17N76W of Jamaica, 1S36E of Suo Piya, birthplace of coffee, 12N41E of Yemen, 35W-74W of Brazil, 5S-35S of Indonesia, birthplace of Mandailing coffee and so on, different figures represent different production place. In additionl, they are also representing different rooms' name, which have shorten the distance in terms of psychology. In space, the sequence has a different heigh, which is full of latitude and longitute gap.

In space axis, beside the coffee making table is a book area (culture). There are some books related to coffee, as well as other pop magazine, design and other information. They are taking as the coffee culture and knowledge to increase the coffee's meaning.

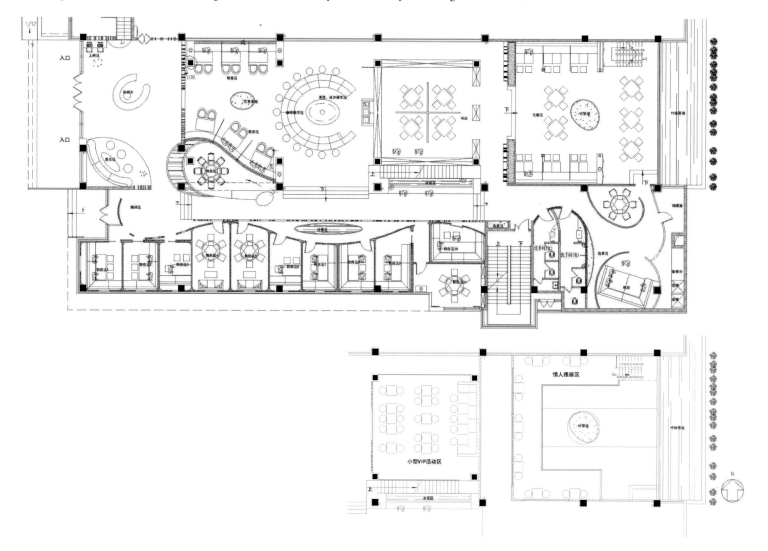

华东 REGION EASTERN CHINA | 301

济南普利售楼处

项目地点：山东济南
设计单位：穆哈地设计咨询
　　　　　（上海）有限公司
　　　　　\MRT DESIGN
设计师：颜呈勋 Bill Yen
建筑面积：1000 m²
项目时间：2011
摄影师：MOSEMAN ELEANOR ELIZABETH

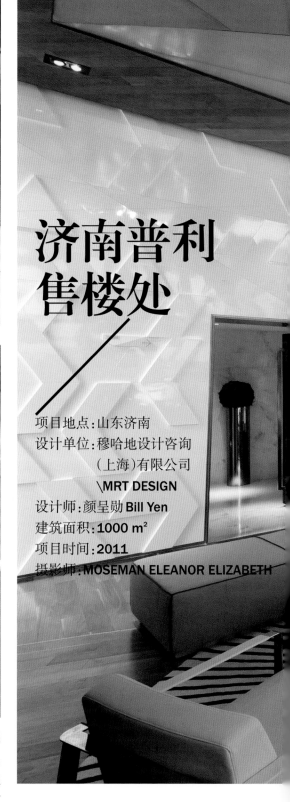

　　济南普利售楼处以白色和木色为主线，木质材料的天然纹理成了本案的装饰，贴近自然，却也简洁婉约。大块面木质材料用金色隔断，打破了原本的突兀，反成了另一种装饰，另一种美。

　　白色墙体上无规则地排列着各种几何造型，繁中透简，乱中有序，白色的主色调选用让整个墙壁看起来纯粹洁净，在自然光及灯光的映衬下，让整个空间看起来更加温暖明亮。

　　本案最大的特点应该是整个空间的倾斜设置，倾斜的装饰，倾斜的天花，倾斜的墙体，这些倾斜元素，营造了一个别样的空间，透着浓浓的探索乐趣；设计师在这里安排了如此美妙的惊喜，给人以无边的想象。

　　交流区域暖色调的墙上运用了大量的钛金板，折射出窗外的美景，随着时间的变化，俨然成了一幅变动的画饰。展示沙盘上方的吊灯是另一亮点，高低错落，呈现出了令人意想不到的演绎结果；再配以本案的主色调与吊灯下方的空间互为呼应，大放光彩。

　　交流区边上两个白色的人物模型，是设计师的心机之处，乍眼看去，还以为是两位售楼人员在迎接宾客。另外，本案选用时尚、简洁的家具，塑造了纯粹的展示交流空间，来此，也是一种享受。

　　楼梯则是另一翻风景，全部选用木材来铺设，木材的天然纹理，延伸了整个空间，营造了特殊的层次造景；这里的楼梯，更像是一个长廊，通往另一个空间，让人充满了探索的欲望。

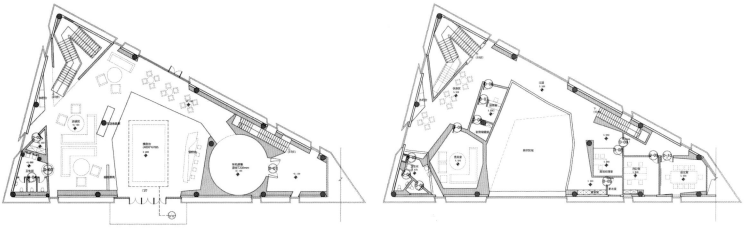

Jinan Puli Sales Office

Jinan Puli Sales Office takes white and wooden color as the tone, the wooden material texture has become the decoration of this case, which is close to nature, simple and graceful. The large wood materials are cut off by gold color, breaking the original oddness, and becoming a kind of decoration, another kind of beauty. There are set various of geometric modeling on the white walls easily, complexity but purity, chaos but orderly. The major tone of whiter color has lighten the wall. Under the natural light and light, it has made the space warmer and brighter.

The biggest feature of this case is the tilt design, tilt decorations, tilt ceiling patterns, tilt wall. These tilt elements have created a different space, showing deep sense of fun. The designer has arranged such a surprise, which would give people a endless imagination.

The warm color wall of communication area adopts much titanium plates, reflecting the beauty of outside. As time goes on, it has become a changable picture. The chandelier suspended on the sand table is another attraction. They are scattered in high and low, showing a unexpected beauty. Coupled with the main color of this case, the chandelier is responding with the space, shining its nice.

The two white figures at the edge of the communication area are set by designer carefully. At first glance, you would think they are the servicers. In addition, this case adopts modern, simple furniture to shape a pure exhibition space. Come here is also a pleasure.

Stairs is another attraction. The stairs adopt wooden materials to pave. The natural texture of wood has extended this entire space and created a special layers of landscape. This stair is more like a long corridor, leading to another space, making people desire to explore.

项目地点：浙江省杭州市
设计单位：杭州意内雅建筑装饰设计有限公司
设计师：朱晓鸣
参与设计：曾文峰、高力勇、赵肖杭
建筑面积：210 m²

杭州西溪 MOHO售楼处

本案展售中心的楼盘，针对的是80后上下从事创意产业为主的时尚群体，结合本案项目所在地空间较为局促等几个方面综合考虑，如何在小场景中创造大印象，如何跳脱房产销售行业同质化令人紧张的交易现场，如何创造一种氛围更容易催化年轻群体的购房欲望，所以就把"她"定位在纯粹、略带童真，甚至添加了几许现代艺术咖啡馆的气氛，做为切入点。

整体的空间采用了极简的双弧线设计，有效地割划了展示区以及内部办公区，模糊化了沙盘区、接洽区和多媒体展示区，使其融合在一起。

空间色彩大面积采用纯净的白色，适当点缀LOGO红色，吻合该项目的视觉形象。

智能感应投影幕的取巧设置，树灯地陈列，还有开放的自由的水吧阅读区，综合传递给每位来访者。轻松而又自由，愉快而又舒畅，畅想而又提前体验优雅小资的未来生活。这也完美表达了MOHO品牌的内在含义：比你想象的更多。

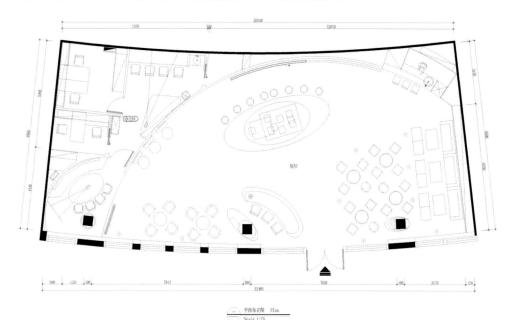

Hangzhou Xixi MOHO Sales Office

This sales office case is created for the fashion group who engage in the innovation industries. Combined with this case that has limited space, there is no much arrangement on how to create a huge impression in such a small space, how to jump out of the tense trading scene of the real estate sales industry, how to create a kind of atmosphere to increase young people's purchase desire. Therefore, we have positioned "her" as a pure and childlike type, even add some modern art café sense as the starting point.

The whole space is adopted the simplest double arc design, which has separated the exhibition area and interior office area effectively, and obscured the Shapan area, negotiation area and multimedia display area and merged them together.

Most of the space is adopted pure white color, dotted with LOGO red, which is in line with the visual image of this project.

The intelligent sensors projection screen is set cleverly. The vertical tree lights, open water room and reading room, all have conveyed to each visitor. Relaxation and freedom, pleasant and comfortable, full of imagination and experience allow you enjoy the elegant life in advance. This also fully expresses the MOHO brand's inherent meaning: much more than your imagination.

千岛湖洲际度假酒店

千岛湖洲际度假酒店拥有255间客房与套房，极具环保主义精神的装修理念提供了健康自然的居住环境，所有客房设计质朴优雅，与当地特色、自然景观融为一体。精致奢华的细节处理和自然野趣的装饰风格，使园林和建筑极具亲和力，充满度假温馨的气息。每间客房及套房空间广阔，拥有超大睡床，可将千岛湖的碧水青山尽收眼底。

三间餐厅与酒廊带来"农家厨房"式的创新美食体验。大堂吧、品茶阁、池畔吧光影浮动，静谧优雅。洲际俱乐部（Club InterContinental）贵宾专属会所、儿童天地（KIDS CLUB）、超大的健身房和铂尊专业水疗品牌的加入，使度假流连忘返，去而不舍。值得一提是，酒店屋顶平台创新设计为天然有机蔬菜种植园，不仅可以观赏还可以品尝；高大的银杏树、艳丽的杨梅林、金灿灿的橘林恍若人间仙境；住店客人可由酒店私有码头直接下湖，尽享东南湖区无敌美景。全新开辟的高端会议场地，18洞观湖高尔夫，环岛豪华游艇，绿色休闲项目和特色水疗给您不一样的自然体验。

洲际会议（InterContinental Meetings）力图捕捉本地文化的精髓，从而激发与会者灵感，令会务活动更加圆满。550 m^2无柱豪华大宴会厅、380 m^2小宴会厅外加4间装修别具匠心，设备先进的会议室可满足各类型高端会议及活动的需求。

项目地点：杭州市淳安县千岛湖镇羡山半岛
设计单位：美国WATG公司
建筑面积：56000 m²
主要材料：天然原木材、石材

InterContinental One Thousand Island Lake Resort is managed by the greatest hotel management group world wide, InterContinental Hotels Group PLC (IHG), It's the unique international luxury brand resort in the Thousand Island Lake.

InterContinental One Thousand Island Lake Resort, located at Xianshan Peninsula, is famous for its Karst Mountains and caves, abundant flora and fauna. The resort, which is the only hotel on the peninsula, offers spectacular panoramic views across the island studded lake and lush green forest, its exclusive yacht is a few minutes walk and the golf club that is a short ride away.

InterContinental One Thousand Island Lake Resort features 255 nature-style rooms and suites with comfortable exquisite interiors and amenities. Three destination restaurants and lounge offer innovative Asian and Western cuisine using produce grown from its own farm/garden. An ideal venue for business and leisure, the resort offers state of the art meeting and banquet space and country adventure and a spa to pamper and rejuvenate your senses.

InterContinental
One Thousand Island Lake Resort

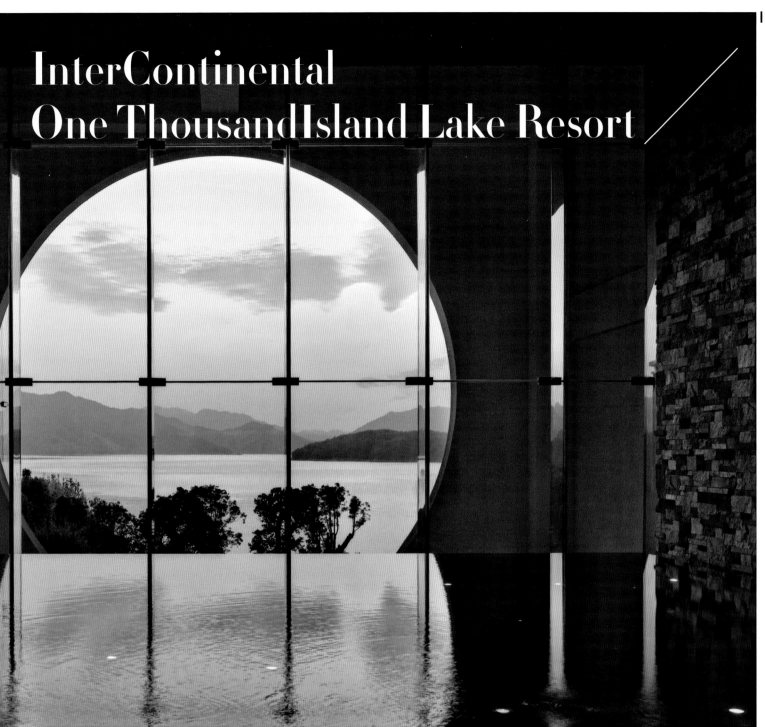

域——中国室内设计年鉴

荣轩茶社

项目地点：浙江台州
设计单位：杭州大相艺术设计有限公司
设计师：蒋建宇、李水、郑小华、楼婷婷、董元军
建筑面积：860 m²
主要材料：青砖、青石板、藤编、松木板染色、
　　　　　铁制灯具、密度板、黑色玄武岩
摄影师：贾方

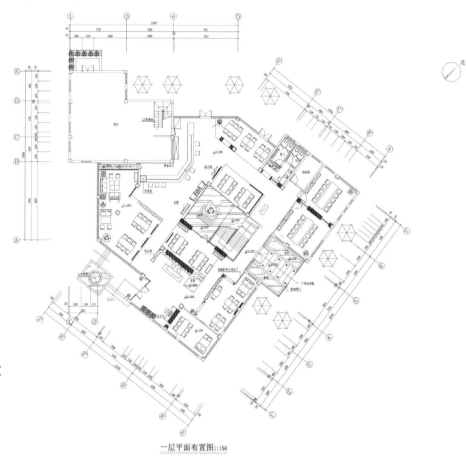

一层平面布置图 1:150

地处台州临海市灵湖公园内，环境清雅宜人。近市区而不喧闹，极具茶禅味之意境。投资人为当地著名张姓美食家，其人品味超绝，行事风格独树一帜。因其好结朋友，圈子广阔，此茶社最初出发点仅在于招待一些兴致相投的喜茶禅之朋友。所以在设计中并无太多商业诉求，力求空间做到心静、远离尘嚣，力图创造一个心灵的净土。设计方在尽可能屏弃元素化的同时尽量减少人工材料的使用，尽可能做到无设计痕迹，并能达到意境上的高远。

本项目在设计中尝试空间特质的全方位体验，通过视觉、听觉、嗅觉、味觉、触觉的综合达到意念上的美妙感观。

Rongxuan Teahouse

Located in the Linghu Park, Linhai city, Taizhou, Rongxuan Teahouse has a comfortable environment. It is close to urban area but has no noise, which has much Buddhist sense. The investor is a well-known local gourmet called Mr. Zhang. He has a high and unique taste, and a special acting style.As he loved to make friends and had broad circle, this teahouse was set for host soem interest-minded friends here at the first time. Therefore, there is no much commercial elements in the design, striving to achieve the quiet and far from the hustle and bustle. The designer tries to minimize elements and reduce the labor and materials. The designer tries to achieve the design has no design traces and reach the lofty sense.

This project has tried the full range experience in the design, through vision, hearing, smell, taste, touch to achieve a wonderful feeling to people.

gxg jeans 宁波城隍庙旗舰店

Gxg jeans品牌男装专为都市青年量身定做，强调年轻人多彩的生活姿态。

店铺设计上，着重从服装风格出发，力求营造自然、清新、优雅的风格，追求激情、创新、自我的气质。将店铺的主题定义为"书"，年轻的一代，有活力，更有内涵。书，是对生活的追求，对品位的追求。因此，店铺从外立面开始，延伸至橱窗，再到室内，充满书的影子，形式纷呈多变：有白色的书模，有真实的外版书，有书的照片，更有书的墙纸、地面、天花，丰富多彩却不夺人眼球。店铺内整面白色书墙，浓重的渲染了学院风格，与服装本身的英伦风悄然呼应。

英伦是服装的主题，因此道具及软装选用上，采用米字国旗加工定做的方式，量身设计了地毯、休息凳、软榻、装饰箱等，配以学院氛围的店铺，时尚、充满活力的感觉呼之欲出。

设计手法上，撇开单一的表现方式，选用MOD手法。MOD，即英文modification（改变、修改）的缩写，将固有的产品进行修改后演变为新型产品，赋予产品新的理念和内涵。比如大门入口即见的长条展示桌，将椅子与桌子镶嵌连为一体，有破又有立，展示桌上既可摆放衣物，也可站立模特，除出样功能外，本身也是一件极好的装饰品，与传统的出样方式大相径庭，更突出gxg jeans品牌特有的创新气质。

项目地点：宁波城隍庙
设计单位：杭州观堂设计
设计师：张健
项目时间：2011.06-2011.10
建筑面积：470 m²
主要材料：地板、书模、铁件
摄影师：王飞
撰文：汤汤

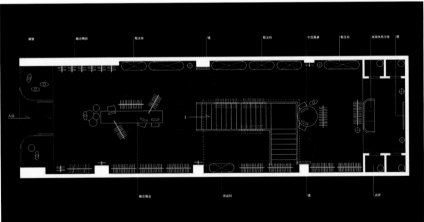

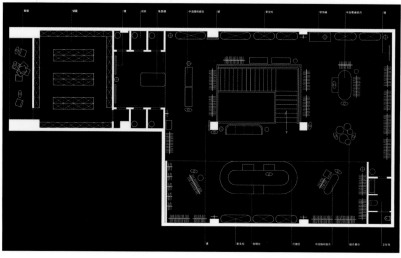

相对于一般店铺收银台只收款的功能，这里特意展示为椭圆形开放式吧台，除了收银电脑，还摆放有打碟机器组，客人可以坐在吧凳上，挑选自己喜欢的CD放进机器自我陶醉，更像是年轻人喜欢的酒吧感觉。

为迎合都市年轻人张扬性格，在二楼特意打造一个空间倒置的区域。天、地、墙都印满了书的图案，在这里，原本应该摆放在桌上的台灯跑到了天花板上；应该站在地上的猪跑到了墙上；应该放在墙上的书，却被踩到了脚底，一切都有些混乱，却带给人全新的体验。

华东 REGION EASTERN CHINA | 333

gxg jeans brand men's clothes

Gxg jeans brand men's clothes is focusing on urban youth, emphasizing the colorful life attitude.

On the shop design, it is focusing on the clothing style, which is striving to create a natural, fresh, elegant style, and pursuing of the passion, creativity, self-temperament. This stop puts "book" as its main theme. It is a younger generation, full of dynamic, meaningful, which is the pursuit of life and the quality. Therefore, from the outer facade of the shope, extending to the window, and then to the indoor, there are full of the book shadows. They are different styles—white book mode, real foreign version book, photos of books, and even wallpapers of book, floors, ceilings and colorful eye-catching scenery. The entire wall white book-wallpaper wall has heavy university style, which is echoing with the British style clothing.

England style is the main theme of clothing. Therefore, on the selecting of the tools and soft equipment, they are adopted customized processing of "Mi" character type flag, customized design carpet, rest chairs, soft chairs, and decorative box and so on. Equipped with the college atmosphere shop, it is full of fashion, vibrant sense. On the design methods, they set aside single performance and adopted MOD tactics. MOD is the abbreviation of the modification (means changer, modify). It modifies the old products into new products, giving the products new concept and connotation. For example, there is a long display table at the entrance, connecting the chairs and desks together. On the display table, it can display the models . At the same time, it is also an excellent decoration, which is different from the traditional style, showing the unique innovation temperament of gxg jeans.

Compared with the common shop, the Cahier is only used for collecting money. In this shop, there is a oval open bar for exhibition. Departing from the cashier computer, there is also placed a DJ Hero. The guests can sit on bars, selecting their favorite CD in the DJ Hero, which is more like a bar favored by young people.

In order to catering the young people, there is set a space-inversion region on the second floor created specially. The sky, earth, and wall are all printed with book patterns. Here, the table that should be placed with lamp are set on the ceiling. The pig that should be on the floor are running on the wall. The book that should be placed on the wall are stepped by our foot. It is a bit mess but new experience.

滕头投资公司生态楼

宁波滕头村被选为联合国最适合人居五百个环境之一，在国内享有一定的政治地位，滕头投资公司生态楼是接待中外领导人，展示滕头村最新成果、动态及办公场所，起到代表滕头形象，设计的重要性绝非等闲。设计主题为生态、环保，展现中国现代农村新气象，造型简约，色彩清新，犹如从田野吹来的一阵清风，使人沐浴在阳光下，心情愉悦。

青山绿水，水波不兴，鹅卵石堆垒的河堤弯弯曲曲，绿色草坪环绕，参天大树相伴，至清至静，生态楼的位置绝佳。原建筑外立面设计得较平淡且不协调，在揭顶时我们开始介入改造，灰绿色为大楼主色调，以盒子的方正形式去叠加，所有的窗往外推做成飘窗，窗的内侧面是白色，远远望去，灰绿色的大楼嵌着一只白色的窗，打破呆板的平面状，现显进出关系。设计还考虑了一些中式元素，在材质上用陶砖百叶与水泥木纹板，在墙面设计了供藤本植物攀爬的架子，时光推移，藤叶会渐渐蔓延至整个墙面，浓密的青翠显现一片生机，四季有不同变化，这是会呼吸的大楼。

建筑的盒子状造型延伸至室内，形成大小不一的方形空间，宛如亭子，水从亭子下面轻轻淌过，鹅卵石铺在水中也堆砌在绿色植物周围，顶上专门设计定做的圆形灯具宛如颗颗晶莹的水珠，人造绿色草坪从地面一直延伸到墙面，大片的白衬着绿，醒目轻松。白色迎接台如一枚鹅卵石，简约的流线型弧状，回收的木板、老船木及用剩的木档用于室内各处，使用自然材料，环保纯朴。大厅有园林的感觉，流曳着江南韵味，同时现代感十足。

办公区域不同部门之间的隔断使用用剩的木档采取叠加方式形成大小不一的方孔，参差落错，局部使用绿灰色陶百叶，富有变化，每个办公室会有攀爬的绿色植物或绿色造景，在这样的环境下办公令人心情舒畅。

展示厅空间不大，整体布局是两个弧线围绕一个球体，形成参观的动线，寓意能源被循环利用，中间的圆球展示的是整个生态系统的模型，地面全部用人工草坪，走在上面如户外散步。顺着参观路线，一个个当地成果陆续展现，视觉在浏览中感触滕头村日新月异的变化。

报告厅暴露的原屋顶被刷成绿色，条形灯盒覆灯光膜，灯盒体量被虚化，光源柔和地被突出，顶部只看到了条形的光带，墙面旧木板跟人工草坪也是条状地进行铺设，它们的间隔有凹凸变化，形成一种节奏关系，加上长条桌子，这些造型以一种平行线的方式进行阵列，报告厅有户外的休闲又不失规整的氛围。

餐厅区域设计主题定位在花园里用餐，那

项目地点: 宁波奉化
设计单位: 宁波市高得装饰设计有限公司
设计师: 范江
参与设计: 唐君
建筑面积: **3850 m²**
主要材料: 陶百叶、水泥木纹板、深色弹性涂料、火烧板、微晶石、高仿真、草坪、回收旧木、老船木、竹地板
摄影师: 潘宇峰
撰文: 洪堃

胃口岂能不好？顶部一片苹果绿，犹如树荫，密度板镂出树叶剪影，刷上白色，绿色与白色的玻璃马赛克以条状铺设从地面延伸至墙面，线条有一种方向感与秩序感，材质晶莹。餐厅淡雅洁净，选用原木色餐桌、餐椅，如童话般浪漫温馨。

大块落地玻璃外就是青翠山谷，峰峦叠起，直接面向大自然，景致揽于心胸，嗟叹环境岂一个"好"字了得！贵宾厅的整个色调略略幽暗，墙面选用深色底一簇簇树叶散落在其中的墙纸上，屏风用镜面不锈钢雕刻了树的剪影，立体含蓄。烟云涌动在平静的湖面是顶部设计的创想，于是密度板镂出云彩状，中间出现镜面玻璃，镜面玻璃倒映着真实的青山绿水，犹如在树丛中看一片纯净的天空。室内室外，风景如画。

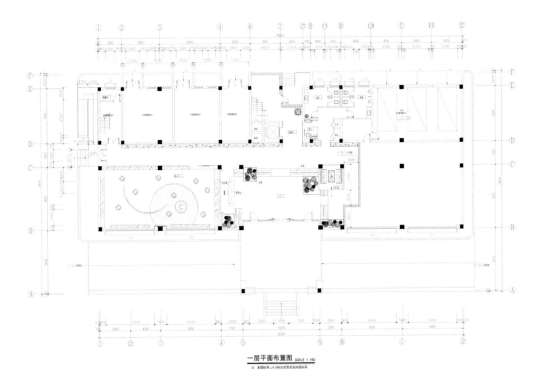

一层平面布置图

Tengtou Investment Company Ecology Building

Tengtou Village in Ningbo city is one of the top 500 suitable places in world selected by the United Nations. It has a certain political status in domestic. Tengtou Investment Company Ecology Building is used to receipt leaders of both home and aboard. The latest achievements, dynamic and office space of Tengtou is on behalf of Tengtou image. The importance of design is by no means very important. The design theme is ecology, environment friendly to show the new image of Chinese modern village. The simple shape, clear color seems like a blow breeze, letting people bathed in the sun.

Green mountain and clean water, the cobblestone bank, green lawns, towering trees are quiet and clear. The ecology building has a perfect location. The original design of building is plain and strange. At the roofing stage, we start to intervene in the transformation. The gray-green color is the major tone. Taking the square shape to be overlapped, all the windows are made into bay window. The inner side of the window is white. Viewing in a far distant way, the gray-green building is embedded with white windows, which has broke the rigid planar, showing an obvious relationship. There are also some Chinese elements in design. We select ceramic louvers tiles as the material, provide shelves on the wall for plant winding. As time going, the leaf will spread to the entire wall gradually, showing a vitality picture. They are changing with four seasons, which we can call it a breathing building.

The box-like building extends to the indoor and forms different square space, seeming like a pavilion. Water flowing on the pavilion softly, the cobbled paved road in the water is also growing many green plants. There is designed custom-made circular lamp specially, which looks like a crystal drop. The artificial green grass is extended from ground to the wall, which is more eye-catching under the white background. The white reception table looks like a cobble stone, it is simple and elegant. The recycled wooden plate, old boat wood, and the wood pieces are used into everywhere in the room. Using the natural materials is environment friendly. The hall has a sense of forest, flowing the South Yangzi River's tone, modern and nice.

The different departments in office area are separated by the rest wood pieces. They are placed in different sizes holes. Partial of them are used gray-green pottery blinds, full of changes. There are climbing green plants or green landscape in each office room. Thus, the office environment looks more comfortable to people.

The exhibition hall space is not large. The overall layout is a ball surrounding by two arcs, forming a moving line for people, which means recycled. The inner ball shows the model of the entire ecosystem. The ground is all paved with artificial grass lawn. Walking on it likes playing outdoors. Walking along the road, the achievements are showing to you one by one, you will enjoy the rapid changes of Tengtou Village when visiting.

The report hall is painted with green. The strip light box is covered with film and the light is highlighted softly. You can only see the light strip at the rop. The old wood plate on wall and the artificial lawn are also paved in strip. Their intervals are changes, forming a sort of rhythm relationship. Coupled with long tables, these shapes are listed in a parallel line. The report hall has an atmosphere of casual and orderly.

The dining area design theme is positioned as garden restaurant. So how can you have a poor appetite? A piece of apple-green is like the shade of a tree. The density plate is hollowed in leaves space. Painting with white color, green and white color glass mosaic will extend from the ground to wall in strip type. The lines have a sense of direction and order, crystal and shinning. The wooden dining table and chair let the restaurant more clear and elegant, liking a place in fairy-tale.

Outside the huge floor window is green valley with many mountains. Facing the nature, you will feel much comfortable. How can a "good" word to cover our feeling!
The whole tone of the VIP room is slightly dark. The wall is dark also. The tree pattern wallpapers on it looks more beautiful. The screen is carved with tree's shadow in three-dimension, showing a humble atmosphere.